T0254940

# Lecture Notes in Computer Science 13611

More information about this series at https://link.springer.com/bookseries/558

Mauricio Reyes · Pedro Henriques Abreu ·
Jaime Cardoso (Eds.)

# Interpretability of Machine Intelligence in Medical Image Computing

5th International Workshop, iMIMIC 2022
Held in Conjunction with MICCAI 2022
Singapore, Singapore, September 22, 2022
Proceedings

*Editors*
Mauricio Reyes ⓘ
University of Bern
Bern, Switzerland

Pedro Henriques Abreu ⓘ
University of Coimbra
Coimbra, Portugal

Jaime Cardoso ⓘ
University of Porto
Porto, Portugal

ISSN 0302-9743                    ISSN 1611-3349 (electronic)
Lecture Notes in Computer Science
ISBN 978-3-031-17975-4           ISBN 978-3-031-17976-1 (eBook)
https://doi.org/10.1007/978-3-031-17976-1

This Springer imprint is published by the registered company Springer Nature Switzerland AG
The registered company address is: Gewerbestrasse 11, 6330 Cham, Switzerland

# Preface

This book constitutes the refereed proceedings of the 5th International Workshop on Interpretability of Machine Intelligence in Medical Image Computing, iMIMIC 2022, held in conjunction with the 25th International Conference on Medical Imaging and Computer-Assisted Intervention, MICCAI 2022.

iMIMIC is a single-track, half-day workshop consisting of high-quality, previously unpublished papers, presented either orally or as a poster, intended to act as a forum for research groups, engineers, and practitioners to present recent algorithmic developments, new results, and promising future directions in interpretability of machine intelligence in medical image computing. Machine learning systems are achieving remarkable performances at the cost of increased complexity. Hence, they are becoming less interpretable, which may cause distrust, potentially limiting clinical acceptance. As these systems are pervasively introduced to critical domains, such as medical image computing and computer assisted intervention, it becomes imperative to develop methodologies allowing insight into their decision making. Such methodologies would help physicians to decide whether they should follow and trust automatic decisions. Additionally, interpretable machine learning methods could facilitate the definition of the legal framework for their clinical deployment. Ultimately, interpretability is closely related to AI safety in healthcare.

This year's iMIMIC was held on September 22, 2022, in Singapore. There was a very positive response to the call for papers for iMIMIC 2022. We received 24 full papers and 10 were accepted for presentation at the workshop, where each paper was reviewed by at least three reviewers in an single-blind process. The accepted papers focus on introducing the challenges and opportunities related to the topic of interpretability of machine learning systems in the context of medical imaging and computer assisted intervention. The high quality of the scientific program of iMIMIC 2022 was due first to the authors who submitted excellent contributions and second to the dedicated collaboration of the international Program Committee and the other researchers who reviewed the papers. We would like to thank all the authors for submitting their contributions and for sharing their research activities. We are particularly indebted to the Program Committee members and to all the reviewers for their precious evaluations, which permitted us to set up this publication. We were also very pleased to benefit from the participation of the invited speakers: Ruth Fong, Princeton University, USA, and Alexander Binder, University of Oslo, Norway. We would like to express our sincere gratitude to these world-renowned experts. Also, we would like to thank our sponsor this year, Varian, a Siemens Healthineers company.

August 2022

Mauricio Reyes
Pedro Henriques Abreu
Jaime Cardoso

# Organization

## General Chairs

| | |
|---|---|
| Mauricio Reyes | University of Bern, Switzerland |
| Jaime Cardoso | INESC TEC and Universidade do Porto, Portugal |
| Jayashree Kalpathy-Cramer | MGH, Harvard University, USA |
| José Amorim | CISUC and University of Coimbra, Portugal |
| Mara Graziani | HES-SO Valais-Wallis, Switzerland |
| Nguyen Le Minh | Japan Advanced Institute of Science and Technology, Japan |
| Pedro Henriques Abreu | CISUC and University of Coimbra, Portugal |
| Wilson Silva | INESC TEC and University of Porto, Portugal |

## Program Committee

| | |
|---|---|
| Alex Bäuerle | Ulm University, Germany |
| André Anjos | Idiap Research Institute, Switzerland |
| Bas H. M. van der Velden | UMC Utrecht, The Netherlands |
| Bettina Finzel | University of Bamberg, Germany |
| Coen de Vente | University of Amsterdam, The Netherlands |
| Cristiano Patrício | Universidade da Beira Interior, Portugal |
| Dwarikanath Mahapatra | Inception Institute of Artificial Intelligence, UAE |
| Eike Petersen | Technical University of Denmark, Denmark |
| Hélder P. Oliveira | INESC TEC, Portugal |
| Helena Montenegro | INESC TEC, Portugal |
| Henning Müller | HES-SO Valais-Wallis, Switzerland |
| Ines Domingues | ISEC, Portugal |
| Isabel Rio-Torto | University of Porto, Portugal |
| Jaime Cardoso (Co-chair) | INESC TEC, Universidade do Porto, Portugal |
| Jana Lipkova | Harvard Medical School, USA |
| John Anderson Garcia Henao | University of Bern, Switzerland |
| Kelwin Fernandes | NILG.AI, Portugal |
| Kerstin Bach | Norwegian University of Science and Technology, Norway |
| Kyi Thar | Mid Sweden University, Sweden |
| Luis Teixeira | INESC TEC, Portugal |
| Mauricio Reyes (Co-chair) | University of Bern, Switzerland |
| Pedro Henriques Abreu (Co-chair) | CISUC and University of Coimbra, Portugal |

Peter Eisert                    Fraunhofer Heinrich Hertz Institute, Germany
Peter Schüffler                 Technical University of Munich, Germany
Plácido L. Vidal                University of A Coruña, Spain
Tiago F. S. Gonçalves           INESC TEC, Portugal
Ute Schmid                      University of Bamberg, Germany

# Contents

# Interpretable Lung Cancer Diagnosis with Nodule Attribute Guidance and Online Model Debugging

Hanxiao Zhang[1], Liang Chen[2], Minghui Zhang[1], Xiao Gu[3], Yulei Qin[4],
Weihao Yu[1], Feng Yao[2], Zhexin Wang[2(✉)], Yun Gu[1,5(✉)],
and Guang-Zhong Yang[1(✉)]

[1] Institute of Medical Robotics, Shanghai Jiao Tong University, Shanghai, China
{geron762,gzyang}@sjtu.edu.cn
[2] Department of Thoracic Surgery, Shanghai Chest Hospital, Shanghai Jiao Tong University, Shanghai, China
wangzhexin001@hotmail.com
[3] Imperial College London, London, UK
[4] Youtu Lab, Tencent, Shanghai, China
[5] Shanghai Center for Brain Science and Brain-Inspired Technology, Shanghai, China

**Abstract.** Accurate nodule labeling and interpretable machine learning are important for lung cancer diagnosis. To circumvent the label ambiguity issue of commonly-used unsure nodule data such as LIDC-IDRI, we constructed a sure nodule data with gold-standard clinical diagnosis. To make the traditional CNN networks interpretable, we propose herewith a novel collaborative model to improve the trustworthiness of lung cancer predictions by self-regulation, which endows the model with the ability to provide explanations in meaningful terms to a human-observer. The proposed collaborative model transfers domain knowledge from unsure data to sure data and encodes a cause-and-effect logic based on nodule segmentation and attributes. Further, we construct a regularization strategy that treats the visual saliency maps (Grad-CAM) not only as post-hoc model interpretation, but also as a rational measure for trustworthy learning in such a way that the CNN features are extracted mainly from intrinsic nodule features. Moreover, similar nodule retrieval makes a nodule diagnosis system more understandable and credible to humans-observers based on the nodule attributes. We demonstrate that the combination of the collaborative model and regularization strategy can provide the best performances on lung cancer prediction and interpretable diagnosis that can automatically: 1) classify the nodule patches; 2) analyse and explain a prediction by nodule segmentation and attributes; and 3) retrieve similar nodules for comparison and diagnosis.

**Keywords:** Lung cancer · Computer-aided diagnosis · Interpretable AI

---

H. Zhang and L. Chen—Joint first authors of this work.

© The Author(s), under exclusive license to Springer Nature Switzerland AG 2022
M. Reyes et al. (Eds.): iMIMIC 2022, LNCS 13611, pp. 1–11, 2022.
https://doi.org/10.1007/978-3-031-17976-1_1

# 1  Introduction

Today's AI systems for CT-based lung cancer diagnosis are highly desirable to gain the trust of clinicians with high-quality data labels and dependable interpretations [6,10,16]. However, based on standard Convolutional Neural Networks (CNNs), most recent approaches [14,20,24,26,27] focus on statistical performance of nodule heterogeneity discrimination within a given nodule dataset LIDC-IDRI [2], instead of model interpretation and generalizability.

Normally, saliency maps [17,31] can retrospectively provide insight and interpret the prediction by highlighting where the model is looking at. However, this cannot explain its predictions in the same way as a human, who can classify objects based on a taxonomy of attributes. This inspired us to design a model which explains its predictions using a set of human-understandable terms. During the annotation of LIDC-IDRI [2,15], nine nodule attributes were assessed by multiple radiologists, which are Subtlety (Sub), Internal Structure (IS), Calcification (Cal), Sphericity (Sph), Margin (Mar), Lobulation (Lob), Spiculation (Spi), Texture (Tex), and Malignancy (Mal). Except for Internal Structure (6 categories) and Calcification (4 categories), each of the attributes is rated on a five-point scale and holds a degree relation (see Fig. 1). Among these attributes, the rating of Malignancy is especially subjective due to the lack of pathologically-proven labels [2]. We term this kind of data as 'unsure(-annotation) data' by its nature of uncertainty. In addition, the outline of each nodule is delineated by multiple radiologists, providing the knowledge of nodule segmentation which, together with nodule attributes, can be considered as understandable concepts for experts to interpret model decisions and make evidence-based diagnoses. This also calls for the need of fair evaluation with a 'sure dataset' that has definite benign-malignant nodule annotations confirmed by pathological examination.

Moreover, saliency maps typically rely on human-experts to examine the corresponding results. By disclosing the salient information of a 'black-box' AI system using interpretable tools, one can intuitively observe some failure cases that the diagnosis model fails to assimilate reliable features from nodule regions (Sect. 4.3 and Fig. 2). These underlying problems are mainly owing to the limitations of deep learning that its model often learns through superficial correlations for data fitting, especially with limited supervision (e.g. patch-level labels) [5]. Due to data scarcity, such circumstance is common yet easily overlooked in medical image analysis [19]. However, saliency maps cannot directly adjust the model if improper regions of attention are highlighted, leading to false and confounding correlations [28]. This encourages us to endow the model with the ability of self-regulation that automatically justifies the feature attention monitored by Grad-CAM [17]. To this end, we use a regularization strategy where Grad-CAM is regarded not only as a post-hoc interpretation, but also as a participant to debug model paired with the reference of nodule segmentation maps.

The feasibility of leveraging Grad-CAM to debug a model has three considerations: 1) it passes the sanity checks to highlighting attentions while many other saliency methods are similar to 'edge detectors' [1,4]; 2) it applies to a wide variety of CNNs for class-discriminative localization [17]; and 3) it is sensitive to the properties of the model parameters, which helps to update model [1].

Further, attribute-based nodule retrieval has the potential to improve the interpretability for lung cancer diagnosis, since it searches for nodules in historically collected data that share similar human-understandable features relative to the one being diagnosed. This mimics the clinical procedure, where clinicians make diagnoses based on their prior knowledge and experience indicated by nodule attributes and segmentation.

The main contribution of this work includes: 1) establishment of a collaborative model for lung cancer prediction guided by the knowledge of nodule segmentation and attributes; 2) introduction of model debugging with Grad-CAM to ensure trustworthiness during training and testing; and 3) provision of interpretable diagnoses for clinicians by attribute-based nodule retrieval.

## 2  Materials

**Unsure Dataset:** According to the practice in [18], we excluded CT scans in LIDC-IDRI [2] with slice thickness larger than 3 mm and selected nodules identified by at least three radiologists. On top of that, we only involve 919 solid nodules (average Texture score = 5). In our work, we do not consider the learning and generating of Internal Structure and Calcification because the inner-classes of these two attributes are extremely imbalanced in this dataset [23]. Accordingly, except for Texture, our work performs the regression of the other six attributes whose average ratings hold sequential degrees. Each nodule segmentation map is generated according to a 50% consensus criterion [13].

**Sure Dataset:** The sure dataset consists of 617 solid nodules (316 benign/301 malignant) collected from 588 patients' CT scans retrospectively in Shanghai Chest Hospital with ethical approval. CT scans in this dataset were acquired by multiple manufacturers where the slice thickness ranges from 0.50 to 3.00 mm with an average of 1.14 ($\pm$ 0.26) mm and the pixel spacing varied from 0.34 to 0.98 mm with an average of 0.60 ($\pm$ 0.22) mm. Each nodule was labeled to a definite class (benign or malignant) confirmed by pathological-proven examination by surgical resection. The exact spatial coordinate and radius of each nodule were annotated by two board-certified radiologists and confirmed by one senior radiologist. In this study, we only include the nodules with a diameter between 3 and 30 mm [3,7]. Note that although there exist some other sure data from NLST trial [21,22], Kaggle's 2017 Data Science Bowl (DSB) competition[1] and LUNGx Challenge dataset [12], we do not include these datasets in our study due to the lack of complete annotations such as position coordinates and pathologic diagnosis.

## 3  Methodology

### 3.1  Collaborative Model Architecture with Attribute-Guidance

In our study, we train a collaborative model (Fig. 1) to jointly conduct nodule segmentation and attribute regression tasks based on the annotation knowledge

---

[1] https://www.kaggle.com/c/data-science-bowl-2017/.

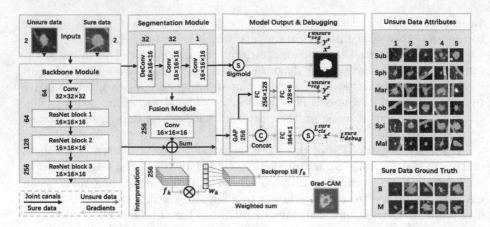

**Fig. 1.** The schematic illustration of the proposed collaborative model for joint learning with sure and unsure data. The basic modules (green bottom color) consist of three parts for feature extraction, nodule segmentation and feature fusion (follow the settings of [28]). In the next stage (yellow), model encodes interactive features for nodule attribute regression and classification, which are regulated with the rational measure of model interpretation (blue). (Color figure online)

of unsure data and perform nodule benign-malignant classification learned from the nodule ground truth of sure data. The proposed collaborative model consists of a backbone for nodule feature extraction, a module for nodule segmentation, a fusion module that combines the features from backbone and segmentation head, and two interactive branches for nodule attribute regression and benign-malignancy classification.

The combined feature maps outputted by the fusion module are fed into the two branches for regression and classification tasks, which act in an interactive way to improve the discriminative ability for nodule prediction by exploring the correlation from attributes to benign-malignant classes. To this end, we first use a fully-connected (FC) layer to generate the intermediate embedding features, and apply another FC layer to output the six attribute scores, which are supervised by unsure data labels. For sure data classification, we first extract the attribute features from the first FC layer of the regression branch, and concatenate these features in the classification branch to make lung cancer prediction.

Different from other works [9,14], we treat the likelihood of Malignancy as a normal attribute rather than the outcome to determine whether a nodule is cancerous or not. This is mainly because: 1) the rating of Malignancy does not have a one-to-one connection with its binary benign-malignant label and retains an uncontrollable subjective bias [29,30]; 2) derived from the experts' knowledge, Malignancy reflects some observable nodule features such as size, shape and brightness; and 3) training six nodule attributes together can implicitly model the internal relationship between them. Such interactive architecture enables more guidance knowledge from nodule segmentation and attributes for sure data to make a decision, although sure data do not have such detailed annotations. We formulate the loss function for the three aforementioned tasks as follows:

$$L_{tasks}^{(un)sure} = \underbrace{g^c log x^c + (1-g^c) log(1-x^c)}_{L_{cls}^{sure}} + \underbrace{1 - \frac{2\sum_i^N y_i^s g_i^s + \theta}{\sum_i^N y_i^s + \sum_i^N g_i^s + \theta}}_{L_{seg}^{unsure}} + \underbrace{\|y^r - g^r\|_2^2}_{L_{reg}^{unsure}} \quad (1)$$

in which, $L_{cls}^{sure}$ is a binary cross-entropy (BCE) loss for the main classification task where $x^c$ is the malignant probability after Sigmoid and $g^c$ is the benign-malignant ground truth of sure data; $L_{seg}^{unsure}$ is a Dice coefficient loss for the auxiliary segmentation task where $y_i^s$ and $g_i^s$ denote the predicted probability and class label of the $i^{th}$ voxel, $N$ is the number of voxels, and $\theta$ is a smoothing coefficient that prevents division by zero; $L_{reg}^{unsure}$ is a mean square error (MSE) loss for the auxiliary attribute regression task where $y^r \in \mathbb{R}^{1 \times n}$ is the regression output, $g^r \in \mathbb{R}^{1 \times n}$ is the average attribute scores rated by radiologists, and $n$ equals to 6 (sub, sph, mar, lob, spi and mal).

## 3.2   Debugging Model with Semantic Interpretation

To deal with the crisis of trustworthiness that happens in the reasoning process of a black-box model, we propose a controllable strategy to constrain the model to **diagnose** 'nodule' rather than arbitrary voxels in the sense of statistics. With the assistance of nodule segmentation map, Grad-CAM [17] is used to interpret and debug model online for trustworthy learning from nodule-relevant features.

Let $f_k (x, y, z)$ represents the unit $k$ at 3D spatial location $(x, y, z)$ of feature maps with length $L$, width $W$ and height $H$ outputted by the fusion module in Fig. 1. To obtain the Grad-CAM, we first compute the gradients of the malignant probability $x^c$ with respect to the feature map $f_k$, $\frac{\partial x^c}{\partial f_k}$. Then, the gradients are global-average-pooled to generate the neuron weights:

$$\omega_k = \frac{1}{L \times W \times H} \sum_x \sum_y \sum_z \frac{\partial x^c}{\partial f_k (x, y, z)} \quad (2)$$

Afterwards, due to using Sigmoid instead of Softmax, we perform a weighted sum of the feature maps $f_k$ to obtain the Grad-CAM map with respect to $x^c$ (benign: $x^c < 0.5$; malignant: $x^c \geqslant 0.5$):

$$Grad\text{-}CAM (x, y, z) = (x^c - 0.5) \sum_k \omega_k f_k (x, y, z) \quad (3)$$

which is then rescaled to [0, 1] by min-max normalization.

To enable trustworthy learning, we regulate the Grad-CAM to concentrate attention on the nodule regions. Guided by the online generated nodule segmentation map, the average Grad-CAM values of nodule regions and background regions can be calculated, which are $Grad\text{-}CAM_{ndl}^{avg} \in [0,1]$, and $Grad\text{-}CAM_{bkg}^{avg} \in [0,1]$. To drive the model to express the features of target object, we enforce $Grad\text{-}CAM_{ndl}^{avg}$ **larger than** $Grad\text{-}CAM_{bkg}^{avg}$, which is formulated as follows:

$$L_{debug}^{sure} = \|x^c - 0.5\|_{l_1} \, max \left\{ 0, Grad\text{-}CAM_{bkg}^{avg} - Grad\text{-}CAM_{ndl}^{avg} + \lambda \right\} \quad (4)$$

where $\lambda$ is a margin parameter (empirically set to 0.5 in this work) and $\|x^c - 0.5\|_{l_1}$ is an adaptive coefficient that encodes the uncertainty of $x^c$ so that model can strengthen the optimization for other tasks if a nodule prediction is of low confidence. In our practical application, we merged the item of $(x^c - 0.5)$ in Eq. (3) and Eq. (4), and made a simplification.

### 3.3   Explanation by Attribute-Based Nodule Retrieval

To enable the interpretable lung cancer diagnosis, we can provide explainability through attribute-based nodule retrieval. Based on the nodule attribute scores $x^r \in \mathbb{R}^{1\times6}$ generated by the collaborative model, we can retrieve K most similar nodules within the historically collected data for the one being diagnosed. The similarity metric used for retrieval is Euclidean Distance. By reading these closely related historical nodule cases, clinicians can acquire more understandable evidence and clues. Meanwhile, the auxiliary attribute scores work as assist-proofs for the diagnosis results and support the user's final decision.

## 4   Experiments and Results

### 4.1   Implementation

In data preprocessing, we first conduct lung segmentation to restrict the valid nodule regions inside the lungs. Then, inspired by the fact that radiologists change CT window widths and centers for nodule diagnosis, we mix lung window [−1000, 400 HU] and mediastinal window [−160, 240 HU] together to generate the nodule inputs. Each window is normalized to the range of [0, 1] and resampled to 0.5 mm/voxel along all three axes using spline interpolation. The final image volume extracted for each nodule is a cube of $64 \times 64 \times 64$ voxels with 2 channels. Data augmentation methods include random flipping, rotation and transposing.

All the experiments are implemented in PyTorch with a single NVIDIA GeForce GTX 1080 Ti GPU and learned using Adam optimizer [11] with the learning rate of 1e−3 (100 epochs). The batch size is set to 1 and group normalization [25] is used after each convolution operation. 5-fold cross-validation is performed, with 20% of the training set used for validation and early stopping.

### 4.2   Quantitative Evaluation

To provide the detailed evaluation of the model performance, we used evaluation metrics including Accuracy, AUC, F1-score, Sensitivity, Specificity, Precision, and Precision$_b$ (Precision in benign class). The results summarized in Table 1

**Table 1.** Quantitative classification performance of comparison methods and ablation study evaluated with sure data by 5-fold cross-validation (threshold $= 0.5$).

| | | Method | Accuracy | AUC | F1-score | Sensitivity | Specificity | Precision | Precision$_b$ |
|---|---|---|---|---|---|---|---|---|---|
| Baselines | 1 | 3D ResNet [8] | 64.03 | 73.26 | 63.09 | 64.05 | 64.00 | 63.63 | 65.97 |
| | 2 | Transfer learning [30] | 67.26 | 73.99 | 65.62 | 64.09 | 70.27 | 67.90 | 67.31 |
| Ablation | 3 | - | 67.29 | 76.28 | 65.85 | 64.40 | 70.03 | 69.00 | 67.24 |
| | 4 | attr | 67.28 | 77.32 | 66.03 | 65.40 | 69.06 | 68.56 | 67.94 |
| | 5 | debug | 69.39 | 76.16 | 67.92 | 66.44 | **72.19** | 69.77 | 69.30 |
| | 6 | attr+debug | 69.20 | 76.57 | 68.86 | 70.74 | 67.74 | 67.98 | 71.49 |
| | 7 | mal+debug (CAM [31]) | 68.24 | 77.29 | 67.22 | 67.39 | 69.04 | 68.69 | 69.56 |
| | 8 | attr+concat | 69.70 | 76.89 | 69.16 | 69.72 | 69.67 | 69.63 | 71.08 |
| | 9 | **attr+concat+debug** | **71.16** | **77.85** | **71.19** | **72.73** | 69.67 | **70.31** | **72.88** |

illustrate the performance of nodule benign-malignancy classification tested on sure data in fair comparison with a normally-used 3D ResNet [8] and a state-of-the-art method [30] which also integrates the knowledge of unsure data. The results show that our best model (the last row) has the ability to predict lung cancer far better than the two baselines, especially for Accuracy, F1-score and Sensitivity. To analyze the impact of each component of our proposed method, we conducted ablation studies in the phase of 'Model Output & Debugging' in Fig. 1 for: (3) only with basic modules; (4) only adding attribute regression (FC: $256 \times 6$); (5) only adding model debugging; (6) without attribute feature concatenation; (7) only adding one attribute ('malignancy', which is the most popular one) for regression and applying CAM [31] for debugging [28]; and (8) without model debugging with Grad-CAM. This shows retaining the single attribute regression or model debugging can barely exceed the performance of 3D ResNet, Transfer learning and the model only with basic modules. The integration of feature concatenation and model debugging plays an important role in improving the performance of nodule benign-malignant discrimination and have a positive effect on reducing overfitting.

### 4.3 Trustworthiness Check and Interpretable Diagnosis

**Trustworthiness Check:** Given the fact that there is no guarantee for a blackbox model to learn nodule-relevant features with respect to model outputs, it is necessary for the human-experts to examine its trustworthiness before considering whether to adopt the model decisions. As illustrated in Fig. 2, the saliency maps (Grad-CAM) of the $1^{st}$ and $2^{nd}$ rows present inexplicable patterns scattered in nodule patches. This implies that 3D ResNet and Transfer learning methods fail our trustworthiness check and can be misleading in real clinical practice. Compared with the $7^{th}$ method (the $3^{rd}$ row), our best method (the $4^{th}$ row) not only appears more effective constraint to extract reliable features in nodule regions, but also has a better quality of nodule segmentation (yellow outline) with a light-weight segmentation module. This benefits from the multi-attribute guidance for nodule discrimination and the superiority of Grad-CAM for model debugging, with their weights being updated by better achieving both

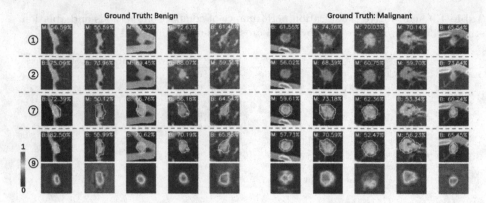

**Fig. 2.** Examples of saliency maps obtained by methods from Table 1. Examples are taken from the central slices of their 3D patches, where the scores are the predicted probabilities to each class and yellow contours denote the nodule segmentation outlines.

nodule segmentation and classification performance. Note that, our method does not completely inhibit feature learning from nodule background according to the last row.

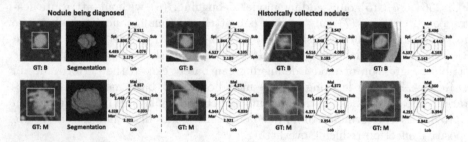

**Fig. 3.** Examples of the attribute-based retrieval for similar nodules (top 3, right part), with respect to the nodule being diagnosed (left part). Attribute scores and 3D segmentation maps are generated by the pre-trained model.

**Interpretable Diagnosis:** Figure 3 shows the examples of attribute-based nodule retrieval using our best model. For the nodule being diagnosed, our system, working as the role of an explainer, can generate its segmentation map and attribute scores, based on which, historically collected nodules with the most similar characteristics can be automatically recalled to support the clinicians to make confident diagnoses.

# 5   Conclusions

Under the fair evaluation of sure data, this paper introduced a new formulation to improve the performance of nodule classification, as well as enhance the trustworthiness of model reasoning and explainability for lung cancer diagnosis. Our superiority mainly comes from the effective cooperation of unsure and sure data knowledge and regulative application of model online debugging with semantic interpretation (Grad-CAM). These innovations empower a diagnosis system more credible and practical during collaboration with clinicians. We believe our formulation can be applied to other classification tasks, where the object segmentation (hand-crafted or automatic) and fine-grained attributes are available to provide regulation for interpretable learning and understandable diagnosis.

**Acknowledgment.** This work was partly supported by Medicine-Engineering Interdisciplinary Research Foundation of Shanghai Jiao Tong University (YG2021QN128), Shanghai Sailing Program (20YF1420800), National Nature Science Foundation of China (No.62003208), Shanghai Municipal of Science and Technology Project (Grant No. 20JC1419500), and Science and Technology Commission of Shanghai Municipality (Grant 20DZ2220400).

# References

1. Adebayo, J., Gilmer, J., Muelly, M., Goodfellow, I., Hardt, M., Kim, B.: Sanity checks for saliency maps. Adv. Neural Inf. Process. Syst. **31**, 1–11 (2018)
2. Armato, S.G., III., et al.: The lung image database consortium (LIDC) and image database resource initiative (IDRI): a completed reference database of lung nodules on CT scans. Med. Phys. **38**(2), 915–931 (2011)
3. Armato, S.G., III., et al.: Lung image database consortium: developing a resource for the medical imaging research community. Radiology **232**(3), 739–748 (2004)
4. Arun, N., et al.: Assessing the trustworthiness of saliency maps for localizing abnormalities in medical imaging. Radiol. Artif. Intell. **3**(6), e200267 (2021)
5. Geirhos, R., et al.: Shortcut learning in deep neural networks. Nat. Mach. Intell. **2**(11), 665–673 (2020)
6. Gunning, D., Stefik, M., Choi, J., Miller, T., Stumpf, S., Yang, G.Z.: XAI-explainable artificial intelligence. Sci. Robot. **4**(37), 1–6 (2019)
7. Hansell, D.M., Bankier, A.A., MacMahon, H., McLoud, T.C., Muller, N.L., Remy, J.: Fleischner society: glossary of terms for thoracic imaging. Radiology **246**(3), 697–722 (2008)
8. He, K., Zhang, X., Ren, S., Sun, J.: Deep residual learning for image recognition. In: Proceedings of the IEEE Conference on Computer Vision and Pattern Recognition, pp. 770–778 (2016)
9. Hussein, S., Cao, K., Song, Q., Bagci, U.: Risk stratification of lung nodules using 3D CNN-based multi-task learning. In: Niethammer, M., et al. (eds.) IPMI 2017. LNCS, vol. 10265, pp. 249–260. Springer, Cham (2017). https://doi.org/10.1007/978-3-319-59050-9_20
10. Jacobs, C., van Ginneken, B.: Google's lung cancer AI: a promising tool that needs further validation. Nat. Rev. Clin. Oncol. **16**(9), 532–533 (2019)

11. Kingma, D.P., Ba, J.: Adam: A method for stochastic optimization. arXiv preprint arXiv:1412.6980 (2014)
12. Kirby, J.S., et al.: LungX challenge for computerized lung nodule classification. J. Med. Imaging **3**(4), 044506 (2016)
13. Kubota, T., Jerebko, A.K., Dewan, M., Salganicoff, M., Krishnan, A.: Segmentation of pulmonary nodules of various densities with morphological approaches and convexity models. Med. Image Anal. **15**(1), 133–154 (2011)
14. Liu, L., Dou, Q., Chen, H., Qin, J., Heng, P.A.: Multi-task deep model with margin ranking loss for lung nodule analysis. IEEE Trans. Med. Imaging **39**(3), 718–728 (2019)
15. McNitt-Gray, M.F., et al.: The lung image database consortium (LIDC) data collection process for nodule detection and annotation. Acad. Radiol. **14**(12), 1464–1474 (2007)
16. Samek, W., Montavon, G., Vedaldi, A., Hansen, L.K., Müller, K.-R.: Explainable AI: Interpreting, Explaining and Visualizing Deep Learning. LNCS (LNAI), vol. 11700. Springer, Cham (2019). https://doi.org/10.1007/978-3-030-28954-6
17. Selvaraju, R.R., Cogswell, M., Das, A., Vedantam, R., Parikh, D., Batra, D.: Gradcam: visual explanations from deep networks via gradient-based localization. In: Proceedings of the IEEE International Conference on Computer Vision, pp. 618–626 (2017)
18. Setio, A.A.A., et al.: Validation, comparison, and combination of algorithms for automatic detection of pulmonary nodules in computed tomography images: the luna16 challenge. Med. Image Anal. **42**, 1–13 (2017)
19. Shad, R., Cunningham, J.P., Ashley, E.A., Langlotz, C.P., Hiesinger, W.: Designing clinically translatable artificial intelligence systems for high-dimensional medical imaging. Nat. Mach. Intell. **3**(11), 929–935 (2021)
20. Shen, W., et al.: Multi-crop convolutional neural networks for lung nodule malignancy suspiciousness classification. Pattern Recogn. **61**, 663–673 (2017)
21. Team, N.L.S.T.R.: The national lung screening trial: overview and study design. Radiology. **258**(1), 243–253 (2011)
22. Team, N.L.S.T.R.: Reduced lung-cancer mortality with low-dose computed tomographic screening. New England J. Med. **365**(5), 395–409 (2011)
23. Wang, Q., et al.: WGAN-based synthetic minority over-sampling technique: improving semantic fine-grained classification for lung nodules in CT images. IEEE Access **7**, 18450–18463 (2019)
24. Wu, B., Zhou, Z., Wang, J., Wang, Y.: Joint learning for pulmonary nodule segmentation, attributes and malignancy prediction. In: 2018 IEEE 15th International Symposium on Biomedical Imaging (ISBI 2018), pp. 1109–1113. IEEE (2018)
25. Wu, Y., He, K.: Group normalization. In: Ferrari, V., Hebert, M., Sminchisescu, C., Weiss, Y. (eds.) ECCV 2018. LNCS, vol. 11217, pp. 3–19. Springer, Cham (2018). https://doi.org/10.1007/978-3-030-01261-8_1
26. Xie, Y., Xia, Y., Zhang, J., Feng, D.D., Fulham, M., Cai, W.: Transferable multimodel ensemble for benign-malignant lung nodule classification on chest CT. In: Descoteaux, M., Maier-Hein, L., Franz, A., Jannin, P., Collins, D.L., Duchesne, S. (eds.) MICCAI 2017. LNCS, vol. 10435, pp. 656–664. Springer, Cham (2017). https://doi.org/10.1007/978-3-319-66179-7_75
27. Xie, Y., et al.: Knowledge-based collaborative deep learning for benign-malignant lung nodule classification on chest CT. IEEE Trans. Med. Imaging **38**(4), 991–1004 (2018)
28. Zhang, H., et al.: Faithful learning with sure data for lung nodule diagnosis. arXiv preprint arXiv:2202.12515 (2022)

29. Zhang, H., et al.: Re-thinking and re-labeling LIDC-IDRI for robust pulmonary cancer prediction. arXiv preprint arXiv:2207.14238 (2022)
30. Zhang, H., Gu, Y., Qin, Y., Yao, F., Yang, G.Z.: Learning with sure data for nodule-level lung cancer prediction. In: Martel, A.L., et al. (eds.) MICCAI 2020. LNCS, vol. 12266, pp. 570–578. Springer, Cham (2020). https://doi.org/10.1007/978-3-030-59725-2_55
31. Zhou, B., Khosla, A., Lapedriza, A., Oliva, A., Torralba, A.: Learning deep features for discriminative localization. In: Proceedings of the IEEE Conference on Computer Vision and Pattern Recognition, pp. 2921–2929 (2016)

# Do Pre-processing and Augmentation Help Explainability? A Multi-seed Analysis for Brain Age Estimation

Daehyun Cho[1] and Christian Wallraven[2(✉)]

[1] Department of Artificial Intelligence, Korea University, Seoul, Korea
`1phantasmas@korea.ac.kr`
[2] Department of Artificial Intelligence and Department of Brain and Cognitive Engineering, Korea University, Seoul, Korea
`christian.wallraven@korea.ac.kr`

**Abstract.** The performance of predicting biological markers from brain scans has rapidly increased over the past years due to the availability of open datasets and efficient deep learning algorithms. There are two concerns with these algorithms, however: they are black-box models, and they can suffer from over-fitting to the training data due to their high capacity. Explainability for visualizing relevant structures aims to address the first issue, whereas data augmentation and pre-processing are used to avoid overfitting and increase generalization performance. In this context, critical open issues are: (i) how robust explainability is across training setups, (ii) how a higher model performance relates to explainability, and (iii) what effects pre-processing and augmentation have on performance and explainability. Here, we use a dataset of 1,452 scans to investigate the effects of augmentation and pre-processing via brain registration on explainability for the task of brain age estimation. Our multi-seed analysis shows that although both augmentation and registration significantly boost loss performance, highlighted brain structures change substantially across training conditions. Our study highlights the need for a careful consideration of training setups in interpreting deep learning outputs in brain analysis.

**Keywords:** Brain age estimation · Deep learning · Explainability · Interpretability · Guided backpropagation

## 1 Introduction

Estimating the age of the brain is essential for detecting abnormalities in brain development, such as neurodegenerative disease or cognitive impairment [20], and has been extensively studied over the past years. As with many other data processing tasks, the advent of deep learning coupled with large, open datasets

**Supplementary Information** The online version contains supplementary material available at https://doi.org/10.1007/978-3-031-17976-1_2.

has significantly increased performance in this domain. These high-performance models, however, suffer from potential overfitting issues given their high capacity and also need to be applied in an explainability framework to open the "black box" [19]. A crucial step for avoiding overfitting has been to use various augmentation strategies that are supposed to increase the robustness and generalizability of the models [30]. Similarly, pre-processing of brain scans, such as anatomical registration, can be done to "help" the models perform better. Explainability methods are then used to create activation maps of those pixels/voxels that are relevant for the model predictions - typically via gradient methods.

We can therefore evaluate deep learning models by their intrinsic metric (i.e., the value of its loss function), but also by their explainability maps (i.e., to what degree do the highlighted regions correspond to known factors of a biological change). In this context it is important to note that training of deep learning models is inherently stochastic due to random weight initialization, dropout, batch size, randomized data augmentations, and stochastic optimization. Hence, running a different "seed" will typically lead to a different sets of weights - even for the same, final loss value. With this in mind, two important open questions remain for gauging the quality of the resulting explanability activations: (i) to what degree are explainability maps consistent across different seeds? and (ii) what effects do pre-processing or augmentation strategies have on explainability?

Here, we present to our knowledge the first, larger-scale study to investigate the effects of seeds, as well as data augmentation and registration in terms of both performance and explainability for the task of brain age prediction. Overall, our contributions are three-fold: first, making use of the stochastic variability across seeds we show that data augmentation results in statistically better (lower) loss compared to non-augmented training for both registered and non-registered brain scans. Second, we investigate the explainability maps via Guided-Backpropagation [31], and find that augmentation results in better-interpretable models, as different seeds share more common voxels. Third, and most importantly, our study uncovers significant changes in explainability already for one deep learning framework across seeds and training setups, highlighting the need for vigilance in interpreting deep learning models for brain age estimation.

## 2    Related Work

Estimation of age from brain scans evolved from "classic" machine learning regressors to current deep neural networks. As raw voxel images were not suitable for the former models to predict age, brain scans were often processed into features first, which were then regressed onto age [2,4,23,33].

The advent of deep learning and the availability of larger brain datasets also changed performance in this task: convolutional neural network architectures showed substantial improvement in age prediction over the past years - both in 2D (analyzing slices of brains, e.g. [13]) or 3D (analyzing the voxels directly, e.g. [3,5,12,18,25]).

In addition to improvements in model architecture, data augmentation, in which the model is fed randomly-transformed scans during training, has been shown to improve performance and to avoid overfitting [30] (see [3] for brain age estimation discussion).

Next, explainability has been added to these black-box models [19], since the prediction of age alone is not sufficient for many applications, as practitioners would also like to know or cross-check which parts of the brain actually are involved in aging. Among the many explainability frameworks, GradCAM [29] and Guided Backpropagation [31] are the most widely-used ones, also in the context of age estimation [12].

Another critical aspect of deep learning models is their stochastic nature: various elements of the training are inherently random, which means that different initializations may lead to different models. Recently, this has led researchers to analyze several, randomly-initialized models in their tasks. This can be done to improve performance (ensembling [18]), but also to gauge the statistical robustness of the results [23]. Here, we take the latter approach and launch an in-depth investigation of explainability across different training setups with seeds used to better analyze the inherent variability of the resultant models.

## 3   Methods

This section describes the dataset and training setups for brain age estimation.

**Table 1.** Left: Dataset demographics. #N = number of scans. Right: age distribution of train (orange) and test (blue) sets. Y-axis denotes density.

| Dataset | #N | Age Mean(std) | Range |
|---------|-----|---------------|-------|
| IXI [11] | 312 | 50.37 (15.932) | 20 - 86 |
| Dallas [24] | 273 | 55.789 (19.478) | 20 - 89 |
| Oasis1 [22] | 315 | 65.698 (9.313) | 18 - 94 |
| Oasis3 [17] | 552 | 54.048 (21.6965) | 42 - 97 |

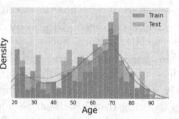

**Dataset:** We used a total of 1,452 brain scans from publicly available datasets (see Table 1). To better gauge generalizability, we kept the same 10% (146 scans) as a *hold-out test* set for all runs. During training, the remaining 90% of the brains, were split again in a 90 train/10 validation set ratio (with varying seeds) to introduce variability for each of 100 fixed random seeds.

**Preprocessing:** All brains were preprocessed starting with skull stripping and normalization through FreeSurfer 6.0 [8]. We then used dipy [7] to register the scans onto the standard MNI152 template including symmetric diffeomorphic registration, removing border voxels outside the brain. All voxels were Min-Max scaled and the scan was cropped to $96 \times 96 \times 96$ voxels.

**Augmentation:** To investigate the effects of augmentation, we chose a set of the three most popular methods: a random mirroring swapped left and right, whereas random affine ($-10$ to $10$) and random elastic deformations (number of control points 7, max displacement 7.5) were used to further increase anatomical variability. For all augmentations, we used the torchio library [26]. During each mini-batch, one augmentation method was selected at random.

**Training Setup:** We used a standard ResNet50 architecture upscaled to 3D [10], with an Adam optimizer with betas 0.9 and 0.999 and a base-learning rate of 0.0001. The loss was a mean squared error. Given the different data domains, the ResNet was non-pretrained. For each experiment configuration, 100 different runs were done on 2 NVIDIA GeForce RTX 2080 Ti via the same 100 random seeds. Early stopping strategy was applied with 20 epochs of patience triggered by validation data. Multiple model checkpoints were saved to trace performance across training. Overall performance for age prediction was only determined on the hold-out test set. All codes are available at https://github.com/1pha/brain-age-prediction.

**Statistical Testing:** To look for the effects of training setups on performance and epochs, we conducted two-sample t-tests across seeds on the hold-out test set. From each seed, we chose the last and best metrics during the training procedure.

**Explainability:** In addition to statistical comparisons, we also used the different seeds to analyze the consistency of resultant explainability maps. Given a trained model checkpoint and one brain scan, we applied Guided-Backpropagation (GBP) [31]. The resultant map of gradients from the mean absolute error (MAE) loss with respect to the feature map was then upsampled to the original input brain scan size, $96 \times 96 \times 96$. Obtaining an average explainability map for one checkpoint started with inferring the 146 brains of the test set, followed by GBP, z-normalizing, and then averaging.

As the only "objective" metric to compare models is the loss metric, we gathered checkpoints from seeds once they reached one of five pre-defined loss thresholds (called "Phases", see Table 2 for threshold values), and aggregated their corresponding explainability maps. We retrieved two quantities from the maps: consistent voxels and highlighted regions. To create the former, as the explainability methods assign a higher value to the voxels that influence the prediction, values lower than the 5% quantile values in the aggregated maps were discarded. In order to determine consistently-contributing voxels for predicting age, we chose those that were implicated in more than half of the seeds. The number of agreeing voxels was selected as an important metric for the robustness of the explainability maps. Next, the chosen top 1% percentile values were than aggregated by regions denoted by the AAL ATLAS [27]. Their results were visualized with nilearn [1].

## 4 Results

### 4.1 Performance

Figure 1(a) shows that augmentation overall results in significant improvements in terms of MAE for both non-registered and registered setups across seed distributions. This improvement is also visualized in Fig. 1(b), which shows the evolution of MAE across training for the different setups. As can be expected, however, augmentation also results in significantly longer training of on average 10 to 13 epochs. We note that prediction performance is somewhat lower compared to other works [18, 25], which is likely due to factors including a slightly smaller dataset as well as less aggressive model optimization. Importantly, however, our main objective was to focus on *relative* differences due to augmentation and pre-processing.

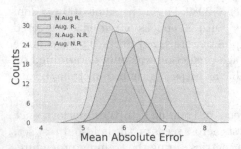
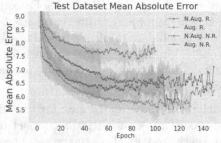

(a) Distribution of best test MAE results for all seeds.

(b) Mean and standard deviation of MAE across epochs.

**Fig. 1.** Mean absolute error statistics on hold-out test dataset. (N.)Aug. = (Non-)Augmented; (N.)R. = (Non-)Registered

## 4.2 Voxel Agreement

We expect that voxels that were repeatedly included in the top-quantiles of the explainability maps across seeds would imply enhanced robustness in explainability. Voxel agreements across conditions are visualized in Fig. 2 - see also Table 2 for detailed values. Here, we find that the augmented-registered condition surpassed the other setups. This seems to be a robust finding also throughout training (Fig. 2), suggesting that this specific training setup identified reliable voxels early.

For both non-registered conditions (blue curves), the number of agreeing voxels was considerably lower compared to the registered conditions (red curves). This is a result that may be expected since models trained with non-aligned brains would find it harder to localize significant regions.

Similarly, when looking at the effects of augmentation, we also found more agreement in general across seeds (compare solid versus dashed lines), indicating that augmentation aids identification of agreeing voxels. Nonetheless, the "effect size" of augmentation on agreement is lower than that of registration.

## 4.3 Atlas-Based Analyses

We next conducted an atlas-based analysis using the top-1% saliency values from each seeds across training procedures and configurations. In this analysis, we chose

**Table 2.** Information on explainability maps for different checkpoints across training setups. Columns denote as follows: #C number of checkpoints reaching given MAE threshold; #E average training epochs for chosen checkpoints; #L average loss for chosen checkpoints.

| Phase | Non-augm./Reg. | | | Augm./Reg. | | |
|---|---|---|---|---|---|---|
| (Threshold) | #C | #E | #L | #C | #E | #L |
| 1 (32) | 100 | 1.76 | 29.75 | 100 | 1.74 | 29.82 |
| 2 (22) | 100 | 2.03 | 22.55 | 100 | 2.02 | 22.84 |
| 3 (7.27) | 100 | 6.13 | 7.24 | 100 | 10.54 | 7.23 |
| 4 (6.0) | 53 | 36.52 | 5.92 | 83 | 36.86 | 5.97 |
| 5 (5.4) | 7 | 53.42 | 5.34 | 26 | 57.03 | 5.35 |
| Phase | Non-augm./Non-reg. | | | Augm./Non-reg. | | |
| (Threshold) | #C | #E | #L | #C | #E | #L |
| 1 (32) | 100 | 1.93 | 30.35 | 100 | 1.89 | 30.45 |
| 2 (22) | 100 | 2.19 | 24.35 | 100 | 2.17 | 23.95 |
| 3 (7.86) | 98 | 16.33 | 7.83 | 99 | 10.78 | 7.83 |
| 4 (6.95) | 15 | 48.66 | 6.92 | 96 | 30.05 | 6.91 |
| 5 (6.1) | 0 | 0 | 0 | 23 | 62.0 | 6.07 |

to aggregate all top-1% values, regarding the variability across seeds as noise to be averaged out. We then averaged the activated voxels in each of the atlas-defined brain regions and ranked these across training setup and phases - see Fig. 3.

For this type of analysis, the main differences were due to registration setups: the registered conditions mostly implicated subcortical regions including CSF, 3rd & lateral ventricles, or parahippocampal regions. In contrast, non-registered models mostly focus on the occipital lobe - cuneus, occipital gyrus and calcarine fissure. Both setups, however, showed high rankings for the brainstem.

## 4.4   Region Validation

To check the alignment between the depicted regions from our models and the brain age literature, here we briefly situate our results in the context of published results both in the brain imaging and the deep learning literature.

One of the consistently-nominated brain region from previous research is the *lateral ventricle* along with the *3rd-ventricle*, most notably due to an increase of width [15,16,32]. This is clearly matching our results in Fig. 3, where these regions for the augmented-registered model have high-rank with minimal deviations.

Similarly we find reports that cerebellar and brainstem volume shrink with age [21], alongside a reduction in thalamic volumes [6,14]. Again, all three regions were implicated in the augmented-registered condition at much higher ranks compared to other training setups.

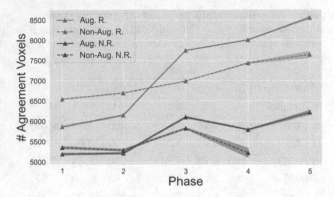

**Fig. 2.** Evolution of agreement across training. The shaded areas indicate the 95% confidence interval across seeds.

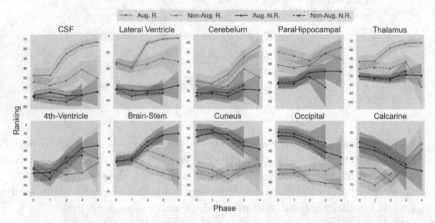

(a) Ranking Charts of RoI saliency value from Guided-Backpropagation.

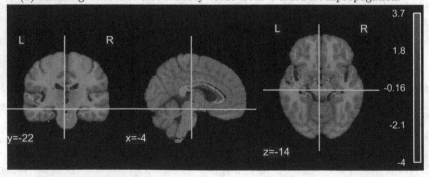

(b) Guided-Backpropagation visualization with augmented-registered condition on the last phase.

**Fig. 3.** Ranking of region of interest from AAL ATLAS across all seeds.

Ranked regions were also compared with those other published studies on brain age prediction with deep learning. [18] and [12] focus on the lateral, 3rd & 4th ventricles and the CSF. The medial temporal structures including the parahippocampals and brain stem are coherent with [28]. [23] mentioned the gray matter cortices, whereas our models mainly focused on the subcorticals and sulci. Overall, significant changes/atrophies in cortical areas could not be found by our work, even with softened thresholds.

# 5   Conclusion

In our work, we showed that pre-processing - a process that tries to minimize anatomical variability for the same voxel - and augmentation - a process that tries to make the model more sensitive to anatomical variability - both improve the performance of brain age prediction. Importantly, we showed this by means of a larger-scale sample that exposed the intrinsic variability of models with different initialization seeds. Going one step further, we used the resultant explainability maps at well-defined, comparable loss thresholds to trace the evolution of predictive brain regions across training phases and training setups. Here, we found that augmentation on registered brains led to the highest agreement across seeds, suggesting the most robust explainability results for this condition. The brain regions implicated in aging for this training setup focused mostly on previously-implicated subcortical brain areas albeit with little to no activation of cortical areas. Overall, our results highlight the need for caution in interpreting bio-markers for brain age prediction (and, similarly, for other tasks) from only one model - a finding that is also visible in the remarkable variability of brain regions implicated in other studies.

Overall, we believe that our study is an important first step in making more replicable and robust statements about bio-markers from brain scans. Future studies will need to generalize these results with larger brain datasets, test additional deep learning architectures (as for example transformers [9]), as well as compare different explainability frameworks.

**Acknowledgements.** This study was supported by the National Research Foundation of Korea under project BK21 FOUR and grants NRF-2017M3C7A1041824, NRF-2019R1A2C2007612, as well as by Institute of Information & Communications Technology Planning & Evaluation (IITP) grants funded by the Korea government (No. 2017-0-00451, Development of BCI based Brain and Cognitive Computing Technology for Recognizing User's Intentions using Deep Learning; No. 2019-0-00079, Department of Artificial Intelligence, Korea University; No. 2021-0-02068, Artificial Intelligence Inovation Hub).

# References

1. Abraham, A., et al.: Machine learning for neuroimaging with scikit-learn. Front. Neuroinform. **14** (2014)

2. Aycheh, H.M., et al.: Biological brain age prediction using cortical thickness data: a large scale cohort study. Front. Aging Neurosci. **10**, 252 (2018)
3. Cole, J.H., et al.: Predicting brain age with deep learning from raw imaging data results in a reliable and heritable biomarker. Neuroimage **163**, 115–124 (2017)
4. Dafflon, J., et al.: An automated machine learning approach to predict brain age from cortical anatomical measures. Hum. Brain Mapp. **41**(13), 3555–3566 (2020)
5. Dinsdale, N.K., et al.: Learning patterns of the ageing brain in MRI using deep convolutional networks. Neuroimage **224**, 117401 (2021)
6. Fama, R., Sullivan, E.V.: Thalamic structures and associated cognitive functions: relations with age and aging. Neurosci. Biobehav. Rev. **54**, 29–37 (2015)
7. Garyfallidis, E., et al.: DIPY, a library for the analysis of diffusion MRI data. Front. Neuroinform. **8**, 8 (2014)
8. Greve, D.N., Fischl, B.: Accurate and robust brain image alignment using boundary-based registration. Neuroimage **48**(1), 63–72 (2009)
9. Gupta, U., Lam, P.K., Ver Steeg, G., Thompson, P.M.: Improved brain age estimation with slice-based set networks. In: 2021 IEEE 18th International Symposium on Biomedical Imaging (ISBI), pp. 840–844. IEEE (2021)
10. Hara, K., Kataoka, H., Satoh, Y.: Can spatiotemporal 3D CNNs retrace the history of 2D CNNs and ImageNet? In: Proceedings of the IEEE Conference on Computer Vision and Pattern Recognition (CVPR), pp. 6546–6555 (2018)
11. Heckemann, R.A., et al.: Information extraction from medical images: developing an e-science application based on the Globus toolkit. In: Proceedings of 2nd UK E-Science Hands Meet (2003)
12. Hepp, T., et al.: Uncertainty estimation and explainability in deep learning-based age estimation of the human brain: results from the German national cohort MRI study. Comput. Med. Imaging Graph. **92**, 101967 (2021)
13. Huang, T.W., et al.: Age estimation from brain MRI images using deep learning. In: 2017 IEEE 14th International Symposium on Biomedical Imaging (ISBI 2017), pp. 849–852. IEEE (2017)
14. Hughes, E.J., et al.: Regional changes in thalamic shape and volume with increasing age. Neuroimage **63**(3), 1134–1142 (2012)
15. Kaye, J.A., DeCarli, C., Luxenberg, J.S., Rapoport, S.I.: The significance of age-related enlargement of the cerebral ventricles in healthy men and women measured by quantitative computed x-ray tomography. J. Am. Geriatr. Soc. **40**(3), 225–231 (1992)
16. Kwon, Y.H., Jang, S.H., Yeo, S.S.: Age-related changes of lateral ventricular width and periventricular white matter in the human brain: a diffusion tensor imaging study. Neural Regen. Res. **9**(9), 986 (2014)
17. LaMontagne, P.J., et al.: Oasis-3: longitudinal neuroimaging, clinical, and cognitive dataset for normal aging and Alzheimer disease. medRxiv (2019)
18. Levakov, G., Rosenthal, G., Shelef, I., Raviv, T.R., Avidan, G.: From a deep learning model back to the brain-identifying regional predictors and their relation to aging. Hum. Brain Mapp. **41**(12), 3235–3252 (2020)
19. Lipton, Z.C.: The mythos of model interpretability: in machine learning, the concept of interpretability is both important and slippery. Queue **16**(3), 31–57 (2018)
20. Lockhart, S.N., DeCarli, C.: Structural imaging measures of brain aging. Neuropsychol. Rev. **24**(3), 271–289 (2014)
21. Luft, A.R., et al.: Patterns of age-related shrinkage in cerebellum and brainstem observed in vivo using three-dimensional MRI Volumetry. Cereb. Cortex **9**(7), 712–721 (1999)

22. Marcus, D.S., Wang, T.H., Parker, J., Csernansky, J.G., Morris, J.C., Buckner, R.L.: Open access series of imaging studies (OASIS): cross-sectional mri data in young, middle aged, nondemented, and demented older adults. J. Cogn. Neurosci. **19**(9), 1498–1507 (2007)
23. Niu, X., Zhang, F., Kounios, J., Liang, H.: Improved prediction of brain age using multimodal neuroimaging data. Hum. Brain Mapp. **41**(6), 1626–1643 (2020)
24. Park, J., et al.: Neural broadening or neural attenuation? Investigating age-related dedifferentiation in the face network in a large lifespan sample. J. Neurosci. **32**(6), 2154–2158 (2012)
25. Peng, H., Gong, W., Beckmann, C.F., Vedaldi, A., Smith, S.M.: Accurate brain age prediction with lightweight deep neural networks. Med. Image Anal. **68**, 101871 (2021)
26. Pérez-García, F., Sparks, R., Ourselin, S.: TorchIO: a python library for efficient loading, preprocessing, augmentation and patch-based sampling of medical images in deep learning. Comput. Methods Progr. Biomed. **208**, 106236 (2021)
27. Rolls, E.T., Huang, C.C., Lin, C.P., Feng, J., Joliot, M.: Automated anatomical labelling atlas 3. Neuroimage **206**, 116189 (2020)
28. Bintsi, K.-M., Baltatzis, V., Hammers, A., Rueckert, D.: Voxel-level importance maps for interpretable brain age estimation. In: Reyes, M., et al. (eds.) IMIMIC/TDA4MedicalData -2021. LNCS, vol. 12929, pp. 65–74. Springer, Cham (2021). https://doi.org/10.1007/978-3-030-87444-5_7
29. Selvaraju, R.R., Cogswell, M., Das, A., Vedantam, R., Parikh, D., Batra, D.: Grad-CAM: visual explanations from deep networks via gradient-based localization. In: Proceedings of the IEEE International Conference on Computer Vision, pp. 618–626 (2017)
30. Shorten, C., Khoshgoftaar, T.M.: A survey on image data augmentation for deep learning. J. Big Data **6**, 60 (2019)
31. Springenberg, J.T., Dosovitskiy, A., Brox, T., Riedmiller, M.A.: Striving for simplicity: the all convolutional net. In: Bengio, Y., LeCun, Y. (eds.) 3rd International Conference on Learning Representations, ICLR 2015, San Diego, CA, USA, May 7–9, 2015, Workshop Track Proceedings (2015)
32. Todd, K.L., et al.: Ventricular and periventricular anomalies in the aging and cognitively impaired brain. Front. Aging Neurosci. **9**, 445 (2018)
33. Wang, B., Pham, T.D.: MRI-based age prediction using hidden Markov models. J. Neurosci. Methods **199**(1), 140–145 (2011)

# Towards Self-explainable Transformers for Cell Classification in Flow Cytometry Data

Florian Kowarsch[1]([✉]), Lisa Weijler[1], Matthias Wödlinger[1,2], Michael Reiter[1,2], Margarita Maurer-Granofszky[2,3], Angela Schumich[2], Elisa O. Sajaroff[4], Stefanie Groeneveld-Krentz[5], Jorge G. Rossi[4], Leonid Karawajew[5], Richard Ratei[6], and Michael N. Dworzak[2,3]

[1] Computer Vision Lab, Faculty of Informatics, TU Wien, Vienna, Austria
fkowarsch@cvl.tuwien.ac.at
[2] Immunological Diagnostics, St. Anna Children's Cancer Research Institute (CCRI), Vienna, Austria
[3] Labdia Labordiagnostik GmbH, Vienna, Austria
[4] Cellular Immunology Laboratory, Hospital de Pediatria "Dr. Juan P. Garrahan", Buenos Aires, Argentina
[5] Department of Pediatric Oncology/Hematology, Charité Universitätsmedizin Berlin, Berlin, Germany
[6] Department of Hematology, Oncology and Tumor Immunology, HELIOS Klinikum Berlin-Buch, Berlin, Germany

**Abstract.** Decisions of automated systems in healthcare can have far-reaching consequences such as delayed or incorrect treatment and thus must be explainable and comprehensible for medical experts. This also applies to the field of automated Flow Cytometry (FCM) data analysis. In leukemic cancer therapy, FCM samples are obtained from the patient's bone marrow to determine the number of remaining leukemic cells. In a manual process, called gating, medical experts draw several polygons among different cell populations on 2D plots in order to hierarchically sub-select and track down cancer cell populations in an FCM sample. Several approaches exist that aim at automating this task. However, predictions of state-of-the-art models for automatic cell-wise classification act as black-boxes and lack the explainability of human-created gating hierarchies. We propose a novel transformer-based approach that classifies cells in FCM data by mimicking the decision process of medical experts. Our network considers all events of a sample at once and predicts the corresponding polygons of the gating hierarchy, thus, producing a verifiable visualization in the same way a human operator does. The proposed model has been evaluated on three publicly available datasets for acute lymphoblastic leukemia (ALL). In experimental comparison it reaches state-of-the-art performance for automated blast cell

---

**Supplementary Information** The online version contains supplementary material available at https://doi.org/10.1007/978-3-031-17976-1_3.

identification while providing transparent results and explainable visualizations for human experts.

**Keywords:** Self-explainable deep learning models · Transformer ·
Flow cytometry gating · Acute lymphoblastic leukemia

# 1  Introduction

Deep Learning models are applicable to a variety of problems arising in healthcare. However, since wrong predictions can have severe consequences, the interpretability of models in this domain is crucial. The output produced by a model needs to be transparent, even for clinicians without any knowledge about the interior of the model. This is also true for the field of automated cell detection in Flow Cytometry (FCM) data. FCM measures the antigen expression levels of blood or bone marrow cells. It is used in research as well as in daily clinical routines for tasks such as immunophenotyping or for monitoring residual numbers of cancer cells (minimal residual disease, MRD) during chemotherapy. A typical sample contains 50–500k cells (also called events) per patient with up to 15 different features (markers) measured. Each feature corresponds to either physical properties of a cell (cell size, granularity) or to the expression level of a specific antigen marker on the cell's surface [18]. While methods for automated MRD assessment already reach human expert level performance [30], they lack interpretability of their predictions. Regardless of a model's performance, clinicians have to manually verify the prediction in a time-consuming process. Using explainable methods could overcome this issue.

Molnar [19] divides existing explainable AI methods into two categories: **Intrinsically interpretable models** are interpretable due to their internal structures. Linear models, decision trees or naive Bayes are common examples of this category. **Post-hoc interpretation methods** analyze a model after training in order to gather explainable insights. Common examples of this category are methods that visualize inner structures of neural networks such as saliency maps [20] and CNN feature visualization techniques or methods, that analyze data input and output pairs of a model to build an explaining description such as LIME [23], shapely values [24,25] and partial dependence plots [9]. In [8] a third category **self-explaining AI** is described, according to which a self-explaining model yields two outputs: a decision and an explanation of that decision.

One way to obtain a self-explaining AI system is to reformulate a prediction task such that the model outputs the same kind of data a domain expert would create to solve or explain a particular problem instance. Instead of directly predicting the solution to a given problem instance, the model is asked to predict a *solution path*. For instance, to solve a linear equation, one can either directly state the solution or provide a series of coherent deductive steps that build an interpretable path to the solution. The latter approach strengthens the trust in

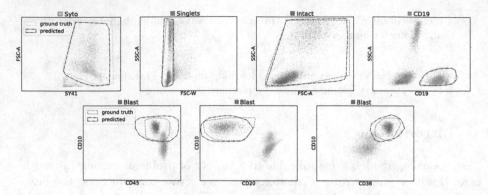

**Fig. 1.** All seven gates of the used gating hierarchy are depicted for an arbitrary FCM sample. Each plot shows a projection of the multidimensional data on two different features. The automated predicted polygons are drawn black and the human operator-created ground truth polygons are drawn in a different color per gate.

the correctness of the solution. While not every data domain admits the modeling of such a solution path, in the field of FCM the **gating hierarchy** can be chosen as an explainable solution path for the problem of cell identification. The conventional procedure to analyze FCM data in the clinical routine is to look at 2D projections of the FCM data and label sub-populations of events by drawing polygons around them [18]. This procedure is called **gating** and the polygons are called **gates**. As illustrated in Fig. 1, gates act as filters by defining the events that are subject to further analysis in other 2D projections (events inside a gate) and the events that will be discarded (events outside the gate). The target population can then be identified by a boolean combination of gates. Gates drawn in specific projections are often applied in sequence, such that one plot only depicts the events selected by the previous plot's polygon. Sequentially applying these gates allows to identify cancer cell populations in the FCM sample. The 2D plots of the data space allow to explicitly depict antigen expressions of the cells in the sample, which are known to be relevant in particular diseases. For example, among other characteristics, CD19 is known to be higher expressed for B-cells [18]. Gating allows analyzing complex patterns of cell populations by a sequence of simpler intermediate steps, which are interpretable by clinicians. Thus, gating is not only a way for finding biologically meaningful sub-populations but has also become the standard for the communication and documentation of FCM sample assessment. Thus it is crucial that the output of machine learning model is compatible with this standard.

In this work we propose a novel method, based on the transformer network, that predicts the polygons of a gating hierarchy to identify cancer cells for MRD assessment in FCM samples of acute lymphoblastic leukemia (ALL) patients.

*Contribution.* This work's contribution is two-fold:

1. A model for ALL blast cell identification is proposed that yields human interpretable visualizations by predicting the polygons of the gating hierarchy while reaching state-of-the-art performance.
2. The proposed model demonstrates how a self-explaining AI systems can be obtained in the medical domain by reformulating the objective function to mimic established human solution procedures.

The remainder of this work is structured as follows. Section 2 gives an overview of methods for automated MRD assessment in FCM data as well as of related architectures. In Sect. 3 the proposed model is described in detail. Section 4 states the conducted experiments, compares the proposed approach to other methods and discusses the results.

## 2   Related Work

Numerous approaches have been established to automate the detection of cell populations in FCM data. The reader is referred to [7] for a more comprehensive review of current trends in automated FCM data analysis. We divide methods for the targeted analysis of FCM data into discriminative and holistic approaches. Approaches that process FCM data event-wise only learn fixed decision regions and are referred to as discriminative approaches. In contrast, holistic approaches process a whole FCM sample and, therefore can account for inter-sample variations, which has been identified as crucial for the correct classification of cell populations with high variability such as leukemic cells [28].

*Discriminative Approaches.* In [1] linear discriminant analysis is proposed for the classification of cell populations as it allows for interpretable performance and reproducibility. Authors in [12] and [10] use a table of marker expression patterns in different cell types as a reference dictionary. Methods based on neural networks include [14, 15].

*Holistic Approaches.* FlowDensity [17] and FlowLearn [16] use an operator's 2D gating strategy as a guideline for detecting cell populations. Recently, a one-class classification approach based on Uniform Manifold Approximation was introduced [29]. Further, Gaussian mixture models (GMM) have proven to be well suited to model cell populations in FCM data [6, 22]. Reiter et al. [22] fit a linear combination of GMMs with labeled components to an unseen sample by Expectation Maximization (EM). [4, 30, 31] are approaches based on neural networks that can process a whole sample at once. Authors in [31] use self-organized maps to obtain a 2D image that a CNN further processes. CellCNN [4] automatically learns a concise cell population representation with a 1D-convolution layer followed by a pooling layer to aggregate information. More recently, Wödlinger et al. [30] presented a method based on the transformer architecture [27] that performs classification on single-cell level, while processing a entire sample in a single neural network forward pass. The attention mechanism of the original transformer architecture [27] entails a quadratic complexity in the input length $\mathcal{O}(n^2)$ of both memory and time, which is unfavorable in the context of FCM data as one sample can contain up to millions of events. Wödlinger et al. thus use the concept of the Induced Set Attention Block (ISAB) as introduced in the set-transformer [13] that reduces the complexity to $\mathcal{O}(n)$.

*Explainable Approaches.* With respect to explainability of results, [10,16,17] can be listed as their results rely on predicted thresholds and hence are interpretable. Algorithmic Population Descriptions (ALPODS), as proposed in [26], is designed to provide explainability by fuzzy reasoning rules in a Bayes decision network expressed in visualizations similar to those generated by domain experts. Another approach related to explainable AI and the method presented in this work is GateFinder [2]. Its goal is to find the shortest yet most discriminative series of 2D polygon gates that lead to a previously specified target population. Although the goal of GateFinder is not targeted analysis, the underlying idea of mimicking the gating strategy of domain experts is similar to the approach presented.

## 3    Methods

The proposed method consists of a trained neural network that is based on the transformer architecture. The model expects a single FCM sample as input, i.e. a set of events $E \in \mathbb{R}^{N \times m}$. $N$ defines the number of events $(50\text{–}500 \times 10^5)$ and $m$ denotes the number of markers (typically 10–20). The network's output are 7 polygons defined by $P = 20$ 2D points each. The polygons describe the gating hierarchy for MRD assessment in ALL data, which implies the cell's class membership. Table 1 displays the predicted gates and the used markers.

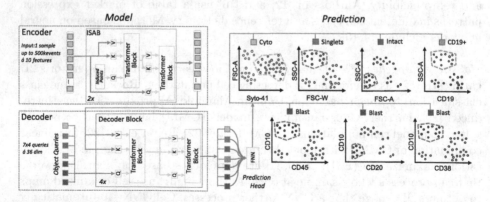

**Fig. 2.** The network architecture consists of the encoder, decoder, prediction head and the resulting polygons that form the gating hierarchy for a given input FCM sample.

**Table 1.** The gates and their used features of the predicted gating hierarchy

| Name | Syto | Singlets | Intact | CD19 | Blast-A | Blast-B | Blast-C |
|---|---|---|---|---|---|---|---|
| Marker y-Axis | FSC-A | SSC-A | SSC-A | SSC-A | CD10 | CD10 | CD10 |
| Marker x-Axis | Syto41 | FSC-W | FSC-A | CD19 | CD45 | CD20 | CD38 |

## 3.1 Architecture

As depicted in Fig. 2, the model's architecture follows an encoder-decoder schema as in [5]. A set-transformer similar to [30] is used for the encoder, consisting of two ISAB blocks. The decoder design is inspired by [5]: for each predicted polygon, four static object queries are learned. The object queries are applied to the encoder's output via cross attention, which is followed by a self-attention layer. Each element of the 7-element long decoder output set is passed through a two-layer fully connected neural network called the prediction head. The resulting 20 2D points per element are used as gate polygon for each of the 7 gates in the ALL gating hierarchy. We empirically evaluated that 20 points are most suitable for the given task. More than 20 points only slightly increase the performance (max 1% median F1-Score) while drastically increasing the network size (see Table 4).

## 3.2 Preprocessing

The operator-annotated polygons comprise two issues regarding their usage as ground truth for training: polygons are typically only roughly estimated, with borders often far away from the nearest events inside the polygon. While this does not affect the effectiveness of the procedure during clinical routine, it introduces a source of ambiguity in the gating process by perturbing the relationship between polygon position and data points. Secondly, for different FCM samples different feature combinations for some of the plots in the gating hierarchy were used by the operator since different operators may use slightly different strategies to track down blast events. However, the model predicts the polygons for a statically predefined set of 2D plot feature combinations. The selected set reflects the most common feature combinations for each gate in the given datasets. We address both issues by computing the convex hull of all events inside the polygon during preprocessing for each gate. The resulting hull serves as adapted training ground truth, which can be created for any required combination of 2D plot features while tightly enclosing the events inside.

## 3.3 Loss Function

$$\mathcal{L}_{poly}(\hat{p}, p) = \sum_i^P \|\hat{p}_{\hat{\sigma}(i)}, p_i\|_1 \text{ with } \hat{\sigma} = \underset{\sigma \in \mathcal{S}_P}{\mathrm{argmin}} \sum_i^P \|p_i, \hat{p}_i\|_1 \qquad (1)$$

The model is trained in a supervised manner. Since the number of polygon vertices differs from sample to sample in the ground truth but is fixed to $P = 20$ for the model prediction, we artificially insert or remove points in the ground truth polygons to obtain $P$ points. Equation 1 states the loss for a predicted polygon $\hat{p}$ where $\hat{\sigma} \in \mathcal{S}_P$ defines a permutation of the polygon points such that every predicted point is matched to one corresponding ground truth point using the Hungarian method [11]. The distance between two points is calculated via L1 norm. Similar to [3,5] we experienced, an auxiliary loss benefits the model convergence. The auxiliary loss performs the same computation as the main loss but after each intermediate layer the following intermediate layers are skipped.

### 3.4   Data Augmentation

To address the low number of training samples (e.g.: $\leq 60$ for the BUE dataset), to overcome inter-laboratory differences and to facilitate learning the relationship between polygon and cell cluster position, four different data augmentation steps are applied to the FCM samples during training: For all events and polygons random linear translations of randomly selected features are applied. For randomly selected gates linear scaling (stretching and squeezing in relation to the center), linear translation and shearing of polygons and their corresponding events are used. Further information is given in the Supplementary material 7.

## 4   Experiments

The same experiments as in [30] have been conducted. In all experiments the proposed model's ability to generalize to new unseen FCM samples (in most cases from different institutes) is tested. The model is implemented in Pytorch 1.10 [21] and trained using the Adam optimizer with a batch size of 12 and a learning rate of $1 \times 10^{-3}$. It consists of 32892 parameters and has been trained on a NVIDIA Gefore RTX 2080 Ti. One model forward pass takes $\approx$400 ms on the used GPU and $\approx$3000 ms on an Intel i7-10750H CPU. Details about the training setup can be found in the provided code on GitHub[1].

### 4.1   Data

The proposed model is evaluated on four different datasets collected across three distinct institutions, measured on three different FCM devices, consisting of over 600 samples in total. From all four datasets, the three datasets VIE14, BLN, BUE are publicly available[2]. All samples have been obtained from the bone marrow of pediatric B-ALL patients on day 15 after induction therapy. The following markers are used in the experiments as they are shared upon all samples: CD10, CD19, CD20, CD34, CD38, CD45 and Syto41 as well as FSC-A, FSC-W and

---

[1] Github Repository.
[2] flowrepository.org.

SSC-A. For a detailed dataset description, the reader is referred to [22] for VIE14, BLN and BUE, and to [30] for VIE20. The experiments have been evaluated by training one network for each dataset.

## 4.2   Results

Table 2 displays the results compared to [22] and [30]. For each experiment the cell classification performance (blast cell vs. non-blast cell) of each sample is summarized with the mean and median F1-Score of all samples in the corresponding test set. The results show that the proposed model is able to reach state-of-the-art performance for blast identification tested on data across different institutes. However, the model under-performs on small training datasets such as BLN and BUE with 70 and 60 training samples. In these cases, the model overfitted during training and was not able to generalize well onto new samples from different sources: Qualitatively inspections revealed that while the cluster positions were mostly correctly predicted, the model failed to predict the correct form of unseen polygon shapes.

**Table 2.** Experiment results of the proposed method compared to GMM [22] and set-transformer [30]. The table reports mean F1-Score/median F1-Score.

| Train | Test | GMM [22] | Transformer [30] | Proposed |
|-------|------|----------|------------------|----------|
| VIE14 | BLN | 0.72/0.81 | 0.77/**0.90** | 0.79/0.88 |
|       | BUE | 0.75/0.90 | 0.82/**0.95** | 0.78/0.89 |
|       | VIE20 | 0.77/0.90 | 0.80/**0.91** | 0.78/0.87 |
| VIE20 | BLN | 0.53/0.58 | 0.68/0.83 | 0.73/**0.85** |
|       | BUE | 0.74/0.88 | 0.75/0.88 | 0.82/**0.92** |
|       | VIE14 | 0.80/0.91 | 0.84/**0.93** | 0.73/0.88 |
| BLN | BUE | 0.65/0.76 | 0.66/**0.87** | 0.69/0.84 |
|     | VIE14 | 0.48/0.48 | 0.82/**0.92** | 0.58/0.73 |
|     | VIE20 | 0.53/0.60 | 0.82/**0.91** | 0.50/0.55 |
| BUE | BLN | 0.62/0.73 | 0.64/**0.78** | 0.57/0.69 |
|     | VIE14 | 0.66/0.73 | 0.83/**0.92** | 0.62/0.69 |
|     | VIE20 | 0.67/0.78 | 0.79/**0.90** | 0.65/0.75 |

The explainable and hierarchical processing of FCM samples in the proposed model elicits two main benefits: first, during model development, unwanted model behaviors such as learned biases can be spotted and addressed. For instance, all applied data augmentation steps were motivated during inspection of the prediction results in the early development stages. Secondly, during inference, the model's prediction can be interpreted. For example, a medical expert can spot and correct a fault in the blast cell classification due to a miss-positioning of a specific polygon in the predicted hierarchy. Take, for example, the CD10CD45-Blast-Gate in Fig. 1: a clinician could adjust the predicted polygon such that no events of the seconded cluster are included in the gate.

## 5   Conclusion

This work proposes a novel transformer-based approach for blast cell detection in FCM samples of ALL patients. The model visually reveals which cells it identifies as blast cells by predicting the polygons of the gating hierarchy for a given FCM sample. This imitates the construction of a gating hierarchy by a human expert in clinical practice and therefore explains why certain events are detected as blast cells. While the proposed model fails to generalize well when trained on small datasets ($\leq$70 samples), its performance is comparable to non-explainable state-of-the-art approaches on more populated datasets ($\geq$180 samples). Future work could address this issue by pretraining the model on artificially generated data. Since the model mimics the decision process of domain experts, it is suitable to be included in the clinical gating routine in the future. The proposed model is designed for pediatric ALL, but the underlying concept could be applied to any disease for which standardized FCM gating hierarchies exist.

**Acknowledgement.** We thank Dieter Printz (FACS Core Unit, CCRI) for flow-cytometer maintenance and quality control, as well as Daniela Scharner and Susanne Suhendra-Chen (CCRI), Jana Hofmann (Charité), Marianne Dunken (HELIOS Klinikum), Marianela Sanz, Andrea Bernasconi, and Raquel Mitchell (Hospital Garrahan) for excellent technical assistance. We are indebted to Melanie Gau, Roxane Licandro, Florian Kleber, Paolo Rota and Guohui Qiao (all from TU Vienna) for valuable contributions to the AutoFLOW project. We thank Markus Kaymer and Michael Kapinsky (both from Beckman Coulter Inc.) for kindly assisting in the provision of customized DuraClone$^{TM}$ tubes for this study as designed by the authors. Notably, Beckman Coulter Inc. did not have any influence on study design, data acquisition and interpretation, or manuscript writing. The study has received funding from the European Union's H2020 Research and Innovation Program through Grant number 825749 "CLOSER: Childhood Leukemia: Overcoming Distance between South America and Europe Regions", the Vienna Business Agency under grant agreement No 2841342 (Project MyeFlow) and by the Marie Curie Industry Academia Partnership & Pathways (FP7-MarieCurie-PEOPLE-2013-IAPP) under grant no. 610872 to project "AutoFLOW" to MND.

## References

1. Abdelaal, T., van Unen, V., Höllt, T., Koning, F., Reinders, M.J., Mahfouz, A.: Predicting cell populations in single cell mass cytometry data. Cytometry A **95**(7), 769–781 (2019)
2. Aghaeepour, N., et al.: GateFinder: projection-based gating strategy optimization for flow and mass cytometry. Bioinformatics **34**(23), 4131–4133 (2018)
3. Al-Rfou, R., Choe, D., Constant, N., Guo, M., Jones, L.: Character-level language modeling with deeper self-attention. In: Proceedings of the AAAI Conference on Artificial Intelligence, vol. 33, pp. 3159–3166 (2019)

4. Arvaniti, E., Claassen, M.: Sensitive detection of rare disease-associated cell subsets via representation learning. Nat. Commun. **8**(14825), 2041–1723 (2017)

5. Carion, N., Massa, F., Synnaeve, G., Usunier, N., Kirillov, A., Zagoruyko, S.: End-to-end object detection with transformers. In: Vedaldi, A., Bischof, H., Brox, T., Frahm, J.-M. (eds.) ECCV 2020. LNCS, vol. 12346, pp. 213–229. Springer, Cham (2020). https://doi.org/10.1007/978-3-030-58452-8_13

6. Chen, X., et al.: Automated flow cytometric analysis across large numbers of samples and cell types. Clin. Immunol. **157**(2), 249–260 (2015)

7. Cheung, M., Campbell, J.J., Whitby, L., Thomas, R.J., Braybrook, J., Petzing, J.: Current trends in flow cytometry automated data analysis software. Cytometry Part A **99**, 1–15 (2021)

8. Elton, D.C.: Self-explaining AI as an alternative to interpretable AI. In: Goertzel, B., Panov, A.I., Potapov, A., Yampolskiy, R. (eds.) AGI 2020. LNCS (LNAI), vol. 12177, pp. 95–106. Springer, Cham (2020). https://doi.org/10.1007/978-3-030-52152-3_10

9. Greenwell, B.M., Boehmke, B.C., McCarthy, A.J.: A simple and effective model-based variable importance measure. arXiv preprint arXiv:1805.04755 (2018)

10. Ji, D., Nalisnick, E., Qian, Y., Scheuermann, R.H., Smyth, P.: Bayesian trees for automated cytometry data analysis. bioRxiv (2018)

11. Kuhn, H.W.: The Hungarian method for the assignment problem. Nav. Res. Logist. Q. **2**(1–2), 83–97 (1955)

12. Lee, H.C., Kosoy, R., Becker, C.E., Dudley, J.T., Kidd, B.A.: Automated cell type discovery and classification through knowledge transfer. Bioinformatics **33**(11), 1689–1695 (2017)

13. Lee, J., Lee, Y., Kim, J., Kosiorek, A., Choi, S., Teh, Y.W.: Set transformer: a framework for attention-based permutation-invariant neural networks. In: International Conference on Machine Learning, pp. 3744–3753. PMLR (2019)

14. Li, H., Shaham, U., Stanton, K.P., Yao, Y., Montgomery, R.R., Kluger, Y.: Gating mass cytometry data by deep learning. Bioinformatics **33**(21), 3423–3430 (2017)

15. Licandro, R., et al.: WGAN latent space embeddings for blast identification in childhood acute myeloid leukaemia. In: 2018 24th International Conference on Pattern Recognition (ICPR), pp. 3868–3873. IEEE (2018)

16. Lux, M., et al.: flowlearn: fast and precise identification and quality checking of cell populations in flow cytometry. Bioinformatics **34**(13), 2245–2253 (2018)

17. Malek, M., Taghiyar, M.J., Chong, L., Finak, G., Gottardo, R., Brinkman, R.R.: flowDensity: reproducing manual gating of flow cytometry data by automated density-based cell population identification. Bioinformatics **31**(4), 606–607 (2014)

18. McKinnon, K.: Flow cytometry: an overview. Curr. Protoc. Immunol. **120**(1), 5-1 (2018)

19. Molnar, C.: Interpretable machine learning. Lulu.com (2020)

20. Nie, W., Zhang, Y., Patel, A.: A theoretical explanation for perplexing behaviors of backpropagation-based visualizations. In: International Conference on Machine Learning, pp. 3809–3818. PMLR (2018)

21. Paszke, A., et al.: Pytorch: an imperative style, high-performance deep learning library. Adv. Neural. Inf. Process. Syst. **32**, 8026–8037 (2019)

22. Reiter, M., et al.: Automated flow cytometric MRD assessment in childhood acute b-lymphoblastic leukemia using supervised machine learning. Cytometry A **95**(9), 966–975 (2019)

23. Ribeiro, M.T., Singh, S., Guestrin, C.: "why should i trust you?" explaining the predictions of any classifier. In: Proceedings of the 22nd ACM SIGKDD International Conference on Knowledge Discovery and Data Mining, pp. 1135–1144 (2016)

24. Shapley, L.S.: A value for n-person games. In: Contributions to the Theory of Games, vol. 2, pp. 307–317 (1953)
25. Sundararajan, M., Najmi, A.: The many shapley values for model explanation. In: International Conference on Machine Learning, pp. 9269–9278. PMLR (2020)
26. Ultsch, A., et al.: An Explainable AI System for the Diagnosis of High Dimensional Biomedical Data. arXiv e-prints arXiv:2107.01820 (2021)
27. Vaswani, A., et al.: Attention is all you need. In: Advances in neural Information Processing Systems, pp. 5998–6008 (2017)
28. Weijler, L., Diem, M., Reiter, M., Maurer-Granofszky, M.: Detecting rare cell populations in flow cytometry data using UMAP. In: 2020 25th International Conference on Pattern Recognition (ICPR), pp. 4903–4909 (2021)
29. Weijler, L., et al.: UMAP based anomaly detection for minimal residual disease quantification within acute myeloid leukemia. Cancers 14(4), 898 (2022)
30. Wodlinger, M., et al.: Automated identification of cell populations in flow cytometry data with transformers. Comput. Biol. Med. 144, 105314 (2022)
31. Zhao, M., et al.: Hematologist-level classification of mature b-cell neoplasm using deep learning on multiparameter flow cytometry data. Cytometry A 97(10), 1073–1080 (2020)

# Reducing Annotation Need
# in Self-explanatory Models for Lung
# Nodule Diagnosis

Jiahao Lu[1,2]($\boxtimes$) (iD), Chong Yin[1,3], Oswin Krause[1], Kenny Erleben[1],
Michael Bachmann Nielsen[2], and Sune Darkner[1]

[1] Department of Computer Science, University of Copenhagen,
Copenhagen, Denmark
lu@di.ku.dk
[2] Department of Diagnostic Radiology, Rigshospitalet,
Copenhagen University Hospital, Copenhagen, Denmark
[3] Department of Computer Science, Hong Kong Baptist University,
Hong Kong, China

**Abstract.** Feature-based self-explanatory methods explain their classification in terms of human-understandable features. In the medical imaging community, this semantic matching of clinical knowledge adds significantly to the trustworthiness of the AI. However, the cost of additional
annotation of features remains a pressing issue. We address this problem
by proposing cRedAnno, a data-/annotation-efficient self-explanatory
approach for lung nodule diagnosis. cRedAnno considerably reduces the
annotation need by introducing self-supervised contrastive learning to
alleviate the burden of learning most parameters from annotation, replacing end-to-end training with two-stage training. When training with hundreds of nodule samples and only 1% of their annotations, cRedAnno
achieves competitive accuracy in predicting malignancy, meanwhile significantly surpassing most previous works in predicting nodule attributes.
Visualisation of the learned space further indicates that the correlation
between the clustering of malignancy and nodule attributes coincides
with clinical knowledge. Our complete code is open-source available:
https://github.com/diku-dk/credanno.

**Keywords:** Explainable AI · Lung nodule diagnosis · Self-explanatory
model · Intrinsic explanation · Self-supervised learning

## 1 Introduction

Lung cancer is one of the leading causes of cancer deaths worldwide due to its
high morbidity and low survival rate [9]. In clinical practice, accurate charac-
terisation of pulmonary nodules in CT images is an essential step for effective
lung cancer screening [28]. Modern deep-learning-based "black box" algorithms,
although achieving accurate classification performance [1], are hardly acceptable
in high-stakes medical diagnosis [26].

M. Reyes et al. (Eds.): iMIMIC 2022, LNCS 13611, pp. 33–43, 2022.
https://doi.org/10.1007/978-3-031-17976-1_4

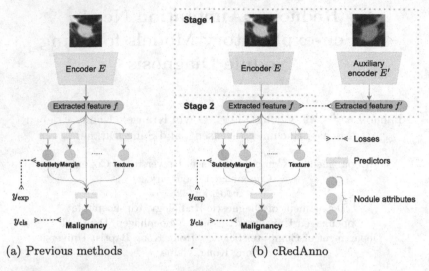

(a) Previous methods                    (b) cRedAnno

**Fig. 1. Concept illustration.** (a) Previous works are trained end-to-end, where all parameters are learned from the annotations. (b) Our proposed cRedAnno uses two-stage training, where most of the parameters are learned during the first stage in a self-supervised manner. Therefore, in the second stage, only few annotations are needed to train the predictors.

Amongst recent efforts to develop explainable AI [4] to bridge this gap [24,26], post-hoc approaches that attempt to explain such "black boxes" are not deemed trustworthy enough [18]. In contrast, feature-based self-explanatory methods are trained to first predict a set of well-known human-interpretable features, and then use these features for the final classification (Fig. 1a) [15,22,23]. This is believed to be especially valuable in medical applications because such semantic matching towards clinical knowledge tremendously increases the AI's trustworthiness [20]. Unfortunately, the required additional annotation on features still limits the applicability of this approach in the medical domain.

This paper aims to minimise additional annotation need for predicting malignancy and nodule attributes in lung CT images. We achieve this by separating the training of model's parameters into two stages, as shown in Fig. 1b. In Stage 1, the majority of parameters are trained using self-supervised contrastive learning [6,11,12] as an encoder to map the input images to a latent space that complies with radiologists' reasoning for nodule malignancy. In Stage 2, a small random portion of labelled samples is used to train a simple predictor for each nodule attribute. Then the predicted human-interpretable nodule attributes are used jointly with the extracted features to make the final classification.

Our experiments on the publicly available LIDC dataset [2] show that with fewer nodule samples and only 1% of their annotations, the proposed approach achieves comparable or better performance compared with state-of-the-art methods using full annotation [7,13,15,16,22], and reaches approximately 90% accuracy in predicting all nodule attributes simultaneously. By visualising the learned space, the extracted features are shown to be highly separable and correlated well with clinical knowledge.

## 2   Method

As the illustrated concept in Fig. 1b, the proposed approach consists of two parts: unsupervised training of the feature encoder and supervised training to predict malignancy with human-interpretable nodule attributes as explanations.

**Unsupervised Feature Extraction.** Due to the outstanding results exhibited by DINO [6], we adopt their framework for unsupervised feature extraction, which trains (i) a primary branch $\{E, H\}_{\theta_{\mathrm{pri}}}$, composed by a feature encoder $E$ and a multi-layer perceptron (MLP) prediction head $H$, parameterised by $\theta_{\mathrm{pri}}$; (ii) an auxiliary branch $\{E, H\}_{\theta_{\mathrm{aux}}}$, which is of the same architecture as the primary branch, while parameterised by $\theta_{\mathrm{aux}}$. After training only the primary encoder $E_{\theta_{\mathrm{pri}}^{E}}$ is used for feature extraction.

The branches are trained using augmented image patches of different scales to grasp the core feature of a sample. For a given input image $x$, different augmented global views $V^g$ and local views $V^l$ are generated [5]: $x \to v \in V^g \cup V^l$. The primary branch is only applied to the global views $v_{\mathrm{pri}} \in V^g$, producing $K$ dimensional outputs $z_{\mathrm{pri}} = E_{\theta_{\mathrm{pri}}^{E}} \circ H_{\theta_{\mathrm{pri}}^{H}} (v_{\mathrm{pri}})$; while the auxiliary branch is applied to all views $v_{\mathrm{aux}} \in V^g \cup V^l$, producing outputs $z_{\mathrm{aux}} = E_{\theta_{\mathrm{aux}}^{E}} \circ H_{\theta_{\mathrm{aux}}^{H}} (v_{\mathrm{aux}})$ to predict $z_{\mathrm{pri}}$. To compute the loss, the output in each branch is passed through a Softmax function scaled by temperature $\tau_{\mathrm{pri}}$ and $\tau_{\mathrm{aux}}$: $p_{\mathrm{aux}} = \mathtt{softmax}(z_{\mathrm{aux}}/\tau_{\mathrm{aux}})$, $p_{\mathrm{pri}} = \mathtt{softmax}((z_{\mathrm{pri}} - \mu)/\tau_{\mathrm{pri}})$, where a bias term $\mu$ is applied to $z_{\mathrm{pri}}$ to avoid collapse [6], and updated at the end of each iteration using the exponential moving average (EMA) of the mean value of a batch with batch size $N$ using momentum factor $\lambda \in [0, 1)$: $\mu \leftarrow \lambda\mu + (1 - \lambda)\frac{1}{N}\sum_{s=1}^{N} z_{\mathrm{pri}}^{(s)}$.

The parameters $\theta_{\mathrm{aux}}$ are learned by minimising the cross-entropy loss between the two branches via back-propagation [12]:

$$\theta_{\mathrm{aux}} \leftarrow \arg\min_{\theta_{\mathrm{aux}}} \sum_{v_{\mathrm{pri}} \in V^g} \sum_{\substack{v_{\mathrm{aux}} \in V^g \cup V^l \\ v_{\mathrm{aux}} \neq v_{\mathrm{pri}}}} \mathcal{L}(p_{\mathrm{pri}}, p_{\mathrm{aux}}), \tag{1}$$

where $\mathcal{L}(p_1, p_2) = -\sum_{c=1}^{C} p_1^{(c)} \log p_2^{(c)}$ for $C$ categories. The parameters $\theta_{\mathrm{pri}}$ of the primary branch are updated by the EMA of the parameters $\theta_{\mathrm{aux}}$ with momentum factor $m \in [0, 1)$:

$$\theta_{\mathrm{pri}} \leftarrow m\theta_{\mathrm{pri}} + (1 - m)\theta_{\mathrm{aux}}. \tag{2}$$

In our implementation, the feature encoders $E$ use Vision Transformer (ViT) [10] as the backbone for their demonstrated ability to learn more generalisable features. Following the basic implementation in DeiT-S [25], our ViTs consist of 12 layers of standard Transformer encoders [27] with 6 attention heads each.

The MLP heads $H$ consist of three linear layers (with GELU activation ) with 2048 hidden dimensions, followed by a bottleneck layer of 256 dimensions, $l_2$ normalisation and a weight-normalised layer [21] to output predictions of $K = 65536$ dimensions, as suggested by [6].

**Supervised Prediction.** After the training of feature encoders is completed, the learned parameters $\theta_{\text{pri}}^{\text{E}}$ in the primary encoder are frozen and all other components are discarded. Given an image $x$ with malignancy annotation $y_{\text{cls}}$ and explanation annotation $y_{\text{exp}}^{(i)}$ for each nodule attribute $i = 1, \cdots, M$, its feature is extracted via the primary encoder: $f = E_{\theta_{\text{pri}}^{\text{E}}}(x)$.

The prediction of each nodule attribute $i$ is generated by a predictor $G_{\text{exp}}^{(i)}$: $z_{\text{exp}}^{(i)} = G_{\text{exp}}^{(i)}(f)$. Then the malignancy prediction $z_{\text{cls}}$ is generated by a predictor $G_{\text{cls}}$ from the concatenation ($\oplus$) of extracted features $f$ and predictions of nodule attributes:

$$z_{\text{cls}} = G_{\text{cls}}(f \oplus z_{\text{exp}}^{(1)} \oplus \cdots \oplus z_{\text{exp}}^{(M)}). \tag{3}$$

The predictors are trained by minimising the cross-entropy loss between the predictions and annotations: $G_{\text{exp}}^{*(i)} = \arg\min \mathcal{L}(y_{\text{exp}}^{(i)}, \text{softmax}(z_{\text{exp}}^{(i)}))$, $G_{\text{cls}}^{*} = \arg\min \mathcal{L}(y_{\text{cls}}, \text{softmax}(z_{\text{cls}}))$.

## 3   Experimental Results

**Data Pre-processing.** We follow the common pre-processing procedure of the LIDC dataset [2] summarised in [3]. Scans with slice thickness larger than 2.5 mm are discarded for being unsuitable for lung cancer screening according to clinical guidelines [14], and the remaining scans are resampled to the resolution of 1 mm$^3$ isotropic voxels. Only nodules annotated by at least three radiologists are retained. Annotations for both malignancy and nodule attributes of each nodule are aggregated by the median value among radiologists. Malignancy score is binarised by a threshold of 3: nodules with median malignancy score larger than 3 are considered malignant, smaller than 3 are considered benign, while the rest are excluded [3]. For each annotation, only a 2D patch of size $32 \times 32\,px$ is extracted from the central axial slice. Although an image is extracted for each annotation, our training(70%)/testing(30%) split is on nodule level to ensure no image of the same nodule exists in both training and testing sets. This results in 276/242 benign/malignant nodules for training and 108/104 benign/malignant nodules for testing.

**Training Settings.** Here we briefly state our training settings and refer to our code repository for further details. The training of the feature extraction follows the suggestions in [6]. The encoders and prediction heads are trained for 300 epochs with an AdamW optimiser and batch size 128, starting from the weights pretrained unsupervisedly on ImageNet [19]. The learning rate is linearly scaled up to 0.00025 during the first 10 epochs and then follows a cosine scheduler to

**Table 1. Prediction accuracy (%) of nodule attributes and malignancy.** The best in each column is **bolded** for full/partial annotation respectively. Dashes (-) denote values not reported by the compared methods. Results of our proposed cRedAnno are highlighted . Observe that cRedAnno in almost all cases outperforms other methods in nodule attributes significantly, and also shows robustness w.r.t. configurations, meanwhile using the fewest nodules and no additional information.

| | Nodule attributes | | | | | | | Malignancy | #nodules | No additional information |
|---|---|---|---|---|---|---|---|---|---|---|
| | Sub | Cal | Sph | Mar | Lob | Spi | Tex | | | |
| Full annotation | | | | | | | | | | |
| HSCNN [22] | 71.90 | 90.80 | 55.20 | 72.50 | - | - | 83.40 | 84.20 | 4252 | ✗c |
| X-Caps [15] | 90.39 | - | 85.44 | 84.14 | 70.69 | 75.23 | 93.10 | 86.39 | 1149 | ✓ |
| MSN-JCN [7] | 70.77 | 94.07 | 68.63 | 78.88 | **94.75** | 93.75 | 89.00 | 87.07 | 2616 | ✗d |
| MTMR [16] | - | - | - | - | - | - | - | **93.50** | 1422 | ✗e |
| cRedAnno (50-NN) | 94.93 | 92.72 | 95.58 | 93.76 | 91.29 | 92.72 | 94.67 | 87.52 | | |
| cRedAnno (250-NN) | **96.36** | 92.59 | 96.23 | 94.15 | 90.90 | 92.33 | 92.72 | 88.95 | 730 | ✓ |
| cRedAnno (trained) | 95.84 | **95.97** | **97.40** | **96.49** | 94.15 | **94.41** | **97.01** | 88.30 | | |
| Partial annotation | | | | | | | | | | |
| WeakSup [13] (1:5a ) | 43.10 | 63.90 | 42.40 | 58.50 | 40.60 | 38.70 | 51.20 | 82.40 | 2558 | ✗f |
| WeakSup [13] (1:3a ) | 66.80 | 91.50 | 66.40 | 79.60 | 74.30 | 81.40 | 82.20 | **89.10** | | |
| cRedAnno (10%b, 50-NN) | 94.93 | 92.07 | **96.75** | **94.28** | **92.59** | 91.16 | **94.15** | 87.13 | | |
| cRedAnno (10%b, 150-NN) | **95.32** | 89.47 | 97.01 | 93.89 | 91.81 | 90.51 | 92.85 | 88.17 | 730 | ✓ |
| cRedAnno (1%b, trained) | 91.81 | **93.37** | 96.49 | 90.77 | 89.73 | **92.33** | 93.76 | 86.09 | | |

[a] $1 : N$ indicates that $\frac{1}{1+N}$ of training samples have annotations on nodule attributes. (All samples have malignancy annotations.)
[b] The proportion of training samples that have annotations on nodule attributes and malignancy.
[c] 3D volume data are used.
[d] Segmentation masks and nodule diameter information are used. Two other traditional methods are used to assist training.
[e] All 2D slices in 3D volumes are used.
[f] Multi-scale 3D volume data are used.

decay till $10^{-6}$. The temperatures for the two branches are set to $\tau_{\mathrm{pri}} = 0.04$, $\tau_{\mathrm{aux}} = 0.1$. The momentum factor $\lambda$ is set to 0.9, while $m$ is increased from 0.996 to 1 following a cosine scheduler. The predictors $G_{\exp}^{(i)}$ and $G_{\mathrm{cls}}$ are jointly trained for 100 epochs with SGD optimisers with momentum 0.9 and batch size 128. The learning rate follows a cosine scheduler with initial value 0.0005 when using full annotation and 0.00025 when using partial annotation.

The data augmentation for encoder training adapts from BYOL [11] and includes multi-crop as in [5]. During the training of the predictors, the input images are augmented following previous works [1,3] on the LIDC dataset.

## 3.1 Prediction Performance of Nodule Attributes and Malignancy

Two categories of experiments are conducted to evaluate the prediction accuracy of both malignancy and each nodule attribute: (i) using k-NN classifiers to assign a label to each feature $f$ extracted from testing images by comparing the dot-product similarity with the ones extracted from training images, without any training; (ii) predicting via trained predictors $G_{\exp}^{(i)}$ and $G_{\mathrm{cls}}$. For simplicity, predictors $G_{\exp}^{(i)}$ and $G_{\mathrm{cls}}$ only use one linear layer. Both k-NN classifier

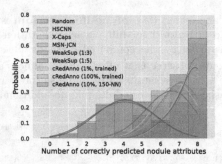

**Fig. 2. Probability of the number of correctly predicted nodule attributes.**
The probabilities of other methods are calculated using their reported prediction accuracy of individual nodule attributes, as in Table 1, where not-reported values are all assumed to be **100% accuracy**. Observe that cRedAnno shows a significantly larger probability of simultaneously predicting all 8 nodule attributes correctly.

and trained predictors are evaluated with full/partial annotation, where partial annotation means only a certain percentage of training samples have annotations on nodule attributes and malignancy. Each annotation is considered independently [22]. The predictions of nodule attributes are considered correct if within $\pm 1$ of aggregated radiologists' annotation [15]. Attribute "internal structure" is excluded from the results because its heavily imbalanced classes are not very informative [7, 13, 15, 16, 22].

The overall prediction performance is summarised in Table 1, comparing with the state-of-the-art. In summary, the results show that our proposed approach can reach simultaneously high accuracy in predicting malignancy and all nodule attributes. This increases the trustworthiness of the model significantly and has not been achieved by previous works. More specifically, when using only 1% annotated samples, our approach achieves comparable or much higher accuracy compared with all previous works in predicting the nodule attributes. Meanwhile, the accuracy of predicting malignancy approaches X-Caps [15] and already exceeds HSCNN [22], which uses 3D volume data. Note that in this case we significantly outperform WeakSup(1:5) [13], which uses 100% malignancy annotations and 16.7% nodule attribute annotations. When using full annotation, our approach outperforms most of the other compared explainable methods in predicting malignancy and all nodule attributes, except "lobulation", where ours is merely worse by absolute 0.6% accuracy. It is worth mentioning that even in this case, we still use the fewest samples: only 518 among the 730 nodules are used for training. In addition, the consistent decent performance also indicates that our approach is reasonably robust w.r.t. to the value $k$ in k-NN classifiers.

To further validate the prediction performance of nodule attributes, for visual clarity, we select 3 representative configurations of our proposed approach and compare them with others in Fig. 2. It can be clearly seen that using our approach, approximately 90% nodules have at least 7 attributes correctly predicted. In contrast, WeakSup(1:5) although reaches over 82.4% accuracy in malignancy prediction, shows no significant difference compared to random guesses in predicting nodule attributes – this shows the opposite of trustworthiness.

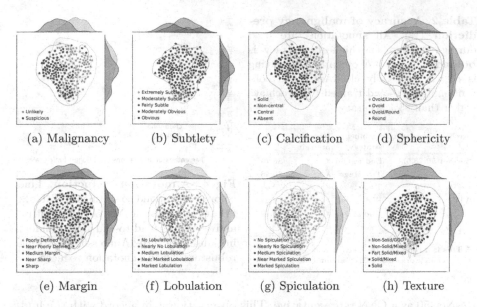

(a) Malignancy     (b) Subtlety     (c) Calcification     (d) Sphericity

(e) Margin     (f) Lobulation     (g) Spiculation     (h) Texture

**Fig. 3. t-SNE visualisation of features extracted from testing images.** Data points are coloured using ground truth annotations. Malignancy shows highly separable in the learned space, and correlates with the clustering in each nodule attribute.

## 3.2 Analysis of Extracted Features in Learned Space

We hypothesise the superior performance of our proposed approach can attribute to the extracted features. So we use t-SNE [17] to further visualise the learned feature. Feature $f$ extracted from each testing image is mapped to a data point in 2D space. Figure 3a to 3h correspond to these data points coloured by the ground truth annotations of malignancy to nodule attribute "texture", respectively. Figure 3a shows that the samples are reasonably linear-separable between the benign/malignant samples even in this dimensionality-reduced 2D space. This provides evidence of our good performance.

Furthermore, the correlation between the nodule attributes and malignancy can be found intuitively in Fig. 3. For example, the cluster in Fig. 3c indicates that solid calcification contributes negatively to malignancy. Similarly, the clusters in Fig. 3e and Fig. 3h indicate that poorly defined margin correlates with non-solid texture, and both of these contribute positively to malignancy. These findings are in accord with the diagnosis process of radiologists [28] and thus further support the trustworthiness of the proposed approach.

## 3.3 Ablation Study

**Validation of Components.** We ablate our proposed approach by comparing with different architectures for encoders $E$, training strategies, and whether to use ImageNet-pretrained weights. The results in Table 2 show that ViT architecture benefits more from the self-supervised contrastive training compared to

**Table 2. Accuracy of malignancy prediction** (%). All annotations are used during training. The highest accuracy is **bolded**. The result of our proposed setting is highlighted. Only cRedAnno and conventional end-to-end trained CNN achieve higher than 85% accuracy.

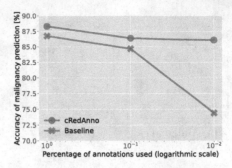

| Arch | #params | Training strategy | ImageNet pretrain | Acc |
|---|---|---|---|---|
| ResNet-50 | 23.5M | End-to-end | ✗ | 86.74* |
| | | Two-stage | ✗ | 70.48 |
| | | Two-stage | ✓ | 70.48 |
| ViT | 21.7M | End-to-end | ✗ | 64.24 |
| | | Two-stage | ✗ | 79.19 |
| | | Two-stage | ✓ | 88.30 |

* This is a representative setting and performance of previous works using CNN architecture.

**Fig. 4. Annotation reduction.** Line colours correspond to settings in Table 2: "Baseline" uses ResNet-50 architecture and is trained end-to-end from random initialisation. cRedAnno shows strong robustness when annotation reduced.

ResNet-50 as a CNN representative. This observation is in accord with the findings in [6,8]. ViT's lowest accuracy in end-to-end training reiterates its requirement for a large amount of training data [10]. Starting from the ImageNet-pretrained weights is also shown to be helpful for ViT but not ResNet-50, probably due to ViT's lack of inductive bias needs far more than hundreds of training samples to compensate [10], especially for medical images. In summary, only the proposed approach and conventional end-to-end training of ResNet-50 achieve higher than 85% accuracy of malignancy prediction.

**Annotation Reduction.** We further plot the malignancy prediction accuracy of the aforementioned winners as the annotations are reduced on a logarithmic scale. As shown in Fig. 4, cRedAnno demonstrates strong robustness w.r.t. annotation reduction. The accuracy of the end-to-end trained ResNet-50 model decreases rapidly to 74.38% when annotations reach only 1%. In contrast, the proposed approach still remains at 86.09% accuracy, meanwhile high accuracy for predicting nodule attributes, as shown in Table 1.

## 4   Conclusion

In this study, we propose cRedAnno to considerably reduce the annotation need in predicting malignancy, meanwhile explaining nodule attributes for lung nodule diagnosis. Our experiments show that even with only 1% annotation, cRedAnno can reach similar or better performance in predicting malignancy compared with state-of-the-art methods using full annotation, and significantly outperforms them in predicting nodule attributes. In addition, our proposed approach is the first to reach over 94% accuracy in predicting all nodule attributes simultaneously. Visualisation of our extracted features provides novel evidence that in the learned space, the clustering of nodule attributes and malignancy is in accord with clinical knowledge of lung nodule diagnosis. Yet the limitations of this approach remain in its generalisability to be validated in other medical image analysis problems.

# References

1. Al-Shabi, M., Lan, B.L., Chan, W.Y., Ng, K.-H., Tan, M.: Lung nodule classification using deep Local–Global networks. Int. J. Comput. Assist. Radiol. Surg. **14**(10), 1815–1819 (2019). https://doi.org/10.1007/s11548-019-01981-7
2. Armato, S.G., et al.: The lung image database consortium (LIDC) and image database resource initiative (IDRI): a completed reference database of lung nodules on CT scans: The LIDC/IDRI thoracic CT database of lung nodules. Med. Phys. **38**(2), 915–931 (2011). https://doi.org/10.1118/1.3528204
3. Baltatzis, V., et al.: The pitfalls of sample selection: a case study on lung nodule classification. In: Rekik, I., Adeli, E., Park, S.H., Schnabel, J. (eds.) PRIME 2021. LNCS, vol. 12928, pp. 201–211. Springer, Cham (2021). https://doi.org/10.1007/978-3-030-87602-9_19
4. Barredo Arrieta, A., et al.: Explainable artificial intelligence (XAI): concepts, taxonomies, opportunities and challenges toward responsible AI. Inf. Fusion **58**, 82–115 (2020). https://doi.org/10.1016/j.inffus.2019.12.012
5. Caron, M., Misra, I., Mairal, J., Goyal, P., Bojanowski, P., Joulin, A.: Unsupervised learning of visual features by contrasting cluster assignments. In: Advances in Neural Information Processing Systems, vol. 33, pp. 9912–9924. Curran Associates, Inc. (2020)
6. Caron, M., et al.: Emerging properties in self-supervised vision transformers. In: Proceedings of the IEEE/CVF International Conference on Computer Vision, pp. 9650–9660 (2021)
7. Chen, W., Wang, Q., Yang, D., Zhang, X., Liu, C., Li, Y.: End-to-End multi-task learning for lung nodule segmentation and diagnosis. In: 2020 25th International Conference on Pattern Recognition (ICPR), pp. 6710–6717. IEEE, Milan, Italy, January 2021. https://doi.org/10.1109/ICPR48806.2021.9412218
8. Chen, X., Xie, S., He, K.: An empirical study of training self-supervised vision transformers. In: 2021 IEEE/CVF International Conference on Computer Vision (ICCV), pp. 9620–9629. IEEE, Montreal, QC, Canada, October 2021. https://doi.org/10.1109/ICCV48922.2021.00950
9. del Ciello, A., et al.: Missed lung cancer: when, where, and why? Diagn. Interv. Radiol. **23**(2), 118–126 (2017). https://doi.org/10.5152/dir.2016.16187
10. Dosovitskiy, A., et al.: An image is worth 16 × 16 words: transformers for image recognition at scale. In: International Conference on Learning Representations, September 2020
11. Grill, J.B., et al.: Bootstrap your own latent - a new approach to self-supervised learning. In: Advances in Neural Information Processing Systems, vol. 33, pp. 21271–21284. Curran Associates, Inc (2020)

12. He, K., Fan, H., Wu, Y., Xie, S., Girshick, R.: Momentum contrast for unsupervised visual representation learning. In: 2020 IEEE/CVF Conference on Computer Vision and Pattern Recognition (CVPR), pp. 9726–9735. IEEE, Seattle, WA, USA, June 2020. https://doi.org/10.1109/CVPR42600.2020.00975

13. Joshi, A., Sivaswamy, J., Joshi, G.D.: Lung nodule malignancy classification with weakly supervised explanation generation. J. Med. Imaging. 8(04), 044502 (2021). https://doi.org/10.1117/1.JMI.8.4.044502

14. Kazerooni, E.A., et al.: ACR–STR practice parameter for the performance and reporting of lung cancer screening thoracic computed tomography (CT): 2014 (Resolution 4)*. J. Thorac. Imaging 29(5), 310–316 (2014). https://doi.org/10.1097/RTI.0000000000000097

15. LaLonde, R., Torigian, D., Bagci, U.: Encoding visual attributes in capsules for explainable medical diagnoses. In: Martel, A.L., et al. (eds.) MICCAI 2020. LNCS, vol. 12261, pp. 294–304. Springer, Cham (2020). https://doi.org/10.1007/978-3-030-59710-8_29

16. Liu, L., Dou, Q., Chen, H., Qin, J., Heng, P.A.: Multi-task deep model with margin ranking loss for lung nodule analysis. IEEE Trans. Med. Imaging 39(3), 718–728 (2020). https://doi.org/10.1109/TMI.2019.2934577

17. van der Maaten, L., Hinton, G.: Visualizing data using t-SNE. J. Mach. Learn. Res. 9(86), 2579–2605 (2008)

18. Rudin, C.: Stop explaining black box machine learning models for high stakes decisions and use interpretable models instead. Nat. Mach. Intell. 1(5), 206–215 (2019). https://doi.org/10.1038/s42256-019-0048-x

19. Russakovsky, O., et al.: ImageNet large scale visual recognition challenge. Int. J. Comput. Vision 115(3), 211–252 (2015). https://doi.org/10.1007/s11263-015-0816-y

20. Salahuddin, Z., Woodruff, H.C., Chatterjee, A., Lambin, P.: Transparency of deep neural networks for medical image analysis: a review of interpretability methods. Comput. Biol. Med. 140, 105111 (2022). https://doi.org/10.1016/j.compbiomed.2021.105111

21. Salimans, T., Kingma, D.P.: Weight normalization: a simple reparameterization to accelerate training of deep neural networks. In: Advances in Neural Information Processing Systems, vol. 29. Curran Associates, Inc. (2016)

22. Shen, S., Han, S.X., Aberle, D.R., Bui, A.A., Hsu, W.: An interpretable deep hierarchical semantic convolutional neural network for lung nodule malignancy classification. Expert Syst. Appl. 128, 84–95 (2019). https://doi.org/10.1016/j.eswa.2019.01.048

23. Stammer, W., Schramowski, P., Kersting, K.: Right for the right concept: revising neuro-symbolic concepts by interacting with their explanations. In: 2021 IEEE/CVF Conference on Computer Vision and Pattern Recognition (CVPR), pp. 3618–3628. IEEE, Nashville, TN, USA, June 2021. https://doi.org/10.1109/CVPR46437.2021.00362

24. Tjoa, E., Guan, C.: A survey on explainable artificial intelligence (XAI): toward medical XAI. IEEE Trans. Neural Netw. Learn. Syst. 32(11), 4793–4813 (2021). https://doi.org/10.1109/TNNLS.2020.3027314

25. Touvron, H., Cord, M., Douze, M., Massa, F., Sablayrolles, A., Jegou, H.: Training data-efficient image transformers & distillation through attention. In: Proceedings of the 38th International Conference on Machine Learning, pp. 10347–10357. PMLR, July 2021

26. van der Velden, B.H., Kuijf, H.J., Gilhuijs, K.G., Viergever, M.A.: Explainable artificial intelligence (XAI) in deep learning-based medical image analysis. Med. Image Anal. **79**, 102470 (2022). https://doi.org/10.1016/j.media.2022.102470
27. Vaswani, A., et al.: Attention is all you need. In: Advances in Neural Information Processing Systems, vol. 30. Curran Associates, Inc. (2017)
28. Vlahos, I., Stefanidis, K., Sheard, S., Nair, A., Sayer, C., Moser, J.: Lung cancer screening: nodule identification and characterization. Transl. Lung Cancer Res. **7**(3), 288–303 (2018). https://doi.org/10.21037/tlcr.2018.05.02

# Attention-Based Interpretable Regression of Gene Expression in Histology

Mara Graziani[1,2(✉)], Niccolò Marini[2], Nicolas Deutschmann[1],
Nikita Janakarajan[1,3], Henning Müller[2], and María Rodríguez Martínez[1]

[1] IBM Research Europe, 8803 Rüschlikon, Switzerland
mara.graziani@hevs.ch
[2] University of Applied Sciences of Western Switzerland, 3960 Sierre, Switzerland
[3] ETH Zürich, 8092 Zürich, Switzerland

**Abstract.** Interpretability of deep learning is widely used to evaluate the reliability of medical imaging models and reduce the risks of inaccurate patient recommendations. For models exceeding human performance, e.g. predicting RNA structure from microscopy images, interpretable modelling can be further used to uncover highly non-trivial patterns which are otherwise imperceptible to the human eye. We show that interpretability can reveal connections between the microscopic appearance of cancer tissue and its gene expression profiling. While exhaustive profiling of all genes from the histology images is still challenging, we estimate the expression values of a well-known subset of genes that is indicative of cancer molecular subtype, survival, and treatment response in colorectal cancer. Our approach successfully identifies meaningful information from the image slides, highlighting hotspots of high gene expression. Our method can help characterise how gene expression shapes tissue morphology and this may be beneficial for patient stratification in the pathology unit. The code is available on GitHub.

**Keywords:** Interpretability · Histopathology · Transcriptomics · Attention

## 1 Introduction

The wide variability of existing interpretability methods, in the form of post-hoc explanations [1,2] or models with ad-hoc transparency constraints [3], has been mainly dedicated to ensuring the safety and reliability of opaque deep neural networks. In medical applications such as digital pathology, saliency maps highlighted the relevance of anomalous nuclei in the detection of tumorous tissue [4], and concept-based analyses confirmed that clinically relevant measures on nuclei area and appearance are learned as intermediate features [5]. Undesired hidden biases and behaviours were detected and corrected to improve model performance [6–8]. Only recently, interpretability techniques were proposed to uncover unknown insights about models with super-human performance, e.g. the algorithm AlphaZero [9] defeating the world chess master. Similarly, deep learning

M. Reyes et al. (Eds.): iMIMIC 2022, LNCS 13611, pp. 44–60, 2022.
https://doi.org/10.1007/978-3-031-17976-1_5

models can predict biomarkers invisible to the human eye [10,11]. For instance, DNA mutations [10] and gene expression profiles were inferred from hematoxylin and eosin (H&E) stained whole slide images (WSIs) [11], demonstrating that complex transcriptomic patterns can be captured from images without the support of sophisticated sequencing machines.

While ensuring patient safety remains the top priority in high-risk clinical environments [12,13], scientific discovery may benefit from unravelling the complex association between RNA expression and tissue microscopy. Identifying the histological patterns that are predictive of gene expression could have an important impact on diagnostic routines, facilitating early patient stratification and aiding the identification of molecular subtypes that are informative of prognosis, survival, and treatment response [14]. For example, clinical trials show that some colorectal (CRC) cancer patients with hypermutated microsatellite instability (MSI) may respond better to immunotherapy than chemotherapy [15].

Our central question is whether we can clarify which regions in CRC H&E slides are the most informative about RNA transcriptomics by using a trainable attention mechanism [16]. It is yet unclear whether the existing algorithms [11,17] infer the bulk average expression of RNA from hotspots of high expression or as a uniform distribution on the entire slide. To obtain interpretable insights, we learn the expression of individual genes rather than the entire transcriptomic profile at once, training several attention-based multiple instance regression models independently. This enables a fine grained analysis of each gene individually and reduces the need for extremely large dataset sizes [11,17] In our results, the attention mechanism brings transparency to the morphological patterns in the tissue that are learned by the model. Hotspots of high expression are highlighted in most of our visualisations and hypotheses can be formulated to relate the histological appearance of tissue to varying gene expression. We show that meaningful information is successfully filtered from the WSIs by the attention, leading to more accurate gene expression estimates and patient stratification. A reduction of 10% in the regression error is seen across genes and patients.

## 2   Methods

### 2.1   Datasets

We use images of colon adenocarcinoma (COAD) and rectal adenocarcinoma (READ) that are publicly available together with matched transcriptomic profiles at The Cancer Genome Atlas (TCGA)[1]. Each biopsy sample is split into three adjacent portions, two of which are used to generate H&E-stained frozen tissue slides, called top-section (TS) and bottom-section (BS), used to verify the presence of sufficient tumour content before sequencing. The central section is used for RNA sequencing. Differently from [11,17], we focus on frozen tissue sections rather than on diagnostic slides, since they are directly adjacent to the

---

[1] https://portal.gdc.cancer.gov, as accessed in June 2022.

sequenced tissue, and thus constitute the best available representation of the RNA profiles [18]. The WSIs are preprocessed with HistoQC [19] to mask out the background and blurred locations.

Gene expression profiles are obtained from the UCSC Xena Browser [20], which links to the Genomic Data Commons [21] version of the TCGA COAD and READ projects. The High Throughput-Sequencing (HT-Seq) raw counts are log2-based Fragments per Kilobase of transcript per Million mapped reads (FPKM) normalised. Because of a distributional shift among institutions, we retain all the patient measurements from a single institution, ignoring duplicates for 23 patients. We focus on the 45 genes in the ColoType signature [22], a gene set that is predictive of CRC prognosis [23] and clinico-pathological variables [24]. The list of genes is in the Appendix Table 1. Single-cell RNA sequencing data are also used for validation[2]. Namely, we use the profiling of 969 single-cells from the CRC resected primary tumours of 11 patients in [25].

As in [26], we exclude patients in preoperative therapy and the rare subtypes of neuroendocrine and signet cell tumours. In total, we use 774 WSIs at 20X magnification from 364 patients. The test set is built by selecting randomly 82 patients and the remaining 282 patients are used for five-fold cross validation.

## 2.2 Multiple Instance Regression of Gene Expression

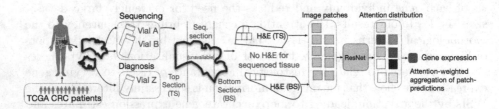

**Fig. 1.** Attention-based Multiple Instance Learning for the semi-transparent regression of RNA gene expression from H&E WSIs.

Our model is an adaptation of the attention-based multiple instance learning approach in [16], which follows the implementation in [27]. We consider a patient $p$ with associated H&E-stained scans of the top and bottom frozen tissue sections $x_t$ and $x_b$, both images in $\mathbb{R}^{w \times h}$, where $w$ is the image width and $h$ is the height. The associated bulk-RNA expression for a given gene $g$ is $y_g \in \mathbb{R}$. To deal with the large input sizes of WSIs, $x_t$ and $x_b$ are represented as a single collection of $N$ non-overlapping adjacent patches obtained by a sliding window $\{x_i\}_{i=0}^{N}$ [28]. The task is to learn a permutation invariant mapping to the real-valued gene label $y_g$ from this joint collection. We assume that each collection contains at least one instance associated with the label [29]. The gene expression $y_g$ can be obtained as a linear-weighted combination of the $y_{i,g}$ predictions for each of the $x_i$ patches,

---

[2] The dataset is available at https://www.weizmann.ac.il/sites/3CA/colorectal.

where the weights are given by the trainable attention. Differently from [16,27], the attention weights $\{a_{i,g}\}_{i=0}^N$ are optimised to predict the continuous gene expression label rather than a binary outcome.

The model, illustrated in Fig. 1, comprises: (i) a convolutional backbone to obtain low-dimensional representations of the input patches, i.e. a ResNet18 [30] pretrained on ImageNet [31]; (ii) a dense layer to predict the gene expression $y_{i,g}$ for each input $x_i$; (iii) an attention network to learn the attention weights.

The output of our model is the expression of a single gene, and we train a different model for each gene in our selection. This is opposed to the approach of the SoA [11], where regression values for over 30000 gene outputs are optimised in a single training. To allow for a fair comparison with the SoA, we adapt their model to output the expression of a single gene at a time. The adapted model predicts the gene expression for each patch, and the predictions are then aggregated by a weighted average with larger weights given to large-valued patch predictions as in [11]. Our trainable attention removes the need for this heuristic.

### 2.3  Attention-Based Model Interpretability

The trainable attention provides transparency on how the model filters the information in the WSIs. Salient regions are identified without the need for an explicit localisation module [16,32]. However, recent debate argued that multiple plausible attention-based explanations may exist and that attention should be interpreted carefully [33,34]. The research in [35] addressed that debate, mentioning that attention weights learned by multiple models should be ensembled by either max or average pooling to reduce the risks of obtaining misleading interpretations. We thus average the attention weights of the models trained on the five folds, ensuring a trustworthy interpretation that focuses only on signal-bearing instances.

### 2.4  Evaluation of Performance and Interpretability

Model performance is evaluated in multiple ways. First, the Median Average Percentage Error (MAPE) of our predictions is compared against the SoA. Both models are also compared to the lower bound of random guessing given by predicting the mean of the training labels. Finally, the significance of our results is verified by computing the non-parametric one-tailed Wilcoxon test between the patient-wise percentage error made by our model and the SoA. Successfully regressed genes are identified as in [11] by evaluating the Pearsons's correlation coefficient $\rho$ between the true and the predicted labels.

The evaluation of model interpretability is non-trivial [17,35]. For morphological patterns that are known to be associated with RNA abnormal expression, e.g. mucinous tissue showing high levels of *MUC2*, we verify that our explanations confirm the existing knowledge. For the newly identified patterns, however, there is little to no ground truth available. In this case, we evaluate how well patch-wise attention weights correlate to the ground truth obtained from analysing single-cell data. We identify, for instance, genes that are co-expressed

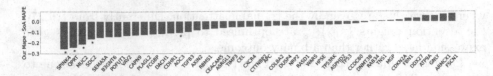

**Fig. 2.** Median difference between the test set errors by our method and the SoA. Statistical significance is shown for p-value <0.05. Lower is better.

in single-cell data according to their Pearson's correlation, and we verify that the patch-wise attentions in overlapping WSI locations correlate for those models. Finally, the benefits of the attention are quantified by quantifying the performance improvement on an auxiliary task, i.e. inferring MSI status. The higher the quality of the attention-based localisation, the more accurate we expect to be the MSI-based stratification of the patients.

## 3    Experiments and Results

### 3.1    Network Training

The models are trained by optimising the mean squared error loss between the predicted gene expression value $\hat{y}_g$ and the label $y_g$. Stochastic gradient descent (SGD) is used with standard hyperparameters (learning rate 0.0001, momentum 0.9, weight decay 0.01) and early stopping (12 epochs patience). The gradient updates affect only the dense layers and the attention mechanism. MSI status is learned in Sect. 3.4 by two additional dense layers trained by SGD minimisation of the weighted binary cross-entropy. For a single model trained on a GPU Tesla V100 for 5 h, we estimate a carbon footprint of 0.65 kgCO$_2$ [36].

### 3.2    Quantitative Model Evaluation

The gene expression estimates obtained by our method are more accurate than the SoA. Across all genes and patients, our model obtains a reduction of 10% in the MAPE, i.e. $0.65 \pm 0.08$ in our model against $0.72 \pm 0.12$ in the SoA. Random guessing as described in Sect. 2.4 achieves MAPE $0.94 \pm 13.0$, showing that both methods are capturing meaningful signals from the WSIs. Computing the Pearson's correlation with the ground-truth labels further confirms the accuracy of our model. *MGP* achieves the highest $\rho$ with 0.71 against 0.69 of the SoA (p-value $< 0.0001$). The predictions for *CCDC80*, *NRP2*, and *RAB34* also show strong correlation with the ground-truth, with $\rho$ 0.65, 0.62 and 0.61 against the SoA $\rho$ at 0.61, 0.61 and 0.55 respectively. The median differences in the errors made by the two models are shown in Fig. 2. The statistical significance of the Wilcoxon test is reported for the individual patient-wise error differences. The detailed results on the full gene set are provided in the Appendix Fig. 6 and Table 2.

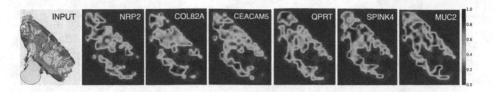

**Fig. 3.** Localised ensemble attention on an unseen test image for six genes.

The largest improvement is observed for *SPINK4*, where the MAPE decreases to $0.62 \pm 1.9$ from $0.88 \pm 3.1$ of the SoA (p-value $< 0.001$). *QPRT, MUC2, AOC3* and *SDC2* report significantly better MAPE than the SoA, achieving respectively $0.74 \pm 1.3$, $0.73 \pm 1.2$, $0.55 \pm 1.3$ and $0.64 \pm 1.4$ against $0.98 \pm 1.1$, $0.95 \pm 1.9$, $0.65 \pm 3.7$ and $0.81 \pm 2.3$ (p-value $< 0.05$). The Pearson's $\rho$ for *QPRT* and *MUC2*, in particular, increase respectively to 0.43 and 0.46 (p-value $< 0.0001$) from 0.35 and 0.13 for the SoA.

### 3.3  Attention-Based Identification of Hotspots and Patterns

Figures 3 and 4 show the predictions and the distribution of the attention for our best gene models on the input WSI. The attention identifies hotspot regions, rather than being uniformly distributed over all patches. Figure 5 visualises more in detail the morphological patterns in the highlighted hotspots of high and low gene expression. We retrieve, for instance, the patches that received the highest normalised attention weights, namely the highest $N_p a_{i,g} \forall i, p$ in the testing set, where $i = 1, \ldots, N_p$, and $N_p$ is the total number of patches for patient $p$[3]. Additional visualisations are in the Appendix and the code repository[4].

Genes that are co-expressed in single-cell data show high attention on the same WSI areas. For example, *NRP2* and *MUC2* show a correlation value of $\rho = 0.40$ at the single-cell gene expression level. Similarly, they display a correlation of 0.63 in the patch-wise attention values. Conversely, *MUC2* is not co-expressed at the single-cell level with *AOC3* and *COL8A2*, and the attention localise on

(a) *NRP2*                                            (b) *MUC2*

**Fig. 4.** Left-to-right order: Original input WSI, raw spatial predictions and attention-weighted predictions. Blue and red show low and high expression respectively. (Color figure online)

---

[3] The normalisation enables the comparison across patients, since $\sum_{i=0}^{N_p} a_{i,g} = 1$.

[4] GitHub link masked for blind submission.

(a) *NRP2*                    (b) *COL8A2*                    (c) *MUC2*

**Fig. 5.** Patches with highest attention for predicted low (blue box) and high (red box) expression. Differing patterns are visible across genes and expression levels. (Color figure online)

different regions, with 0.16 and 0.12 patch-wise correlations respectively. The results for all gene couples are in the Appendix Fig. 10.

### 3.4  Quantitative Evaluation of the Attention

The MSI status is inferred from the gene expression values regressed from the H&E slides. The input noise is reduced by selecting only the genes that were predicted with Pearson's p-value $< 0.05$ by each method. This leads to 37 genes with our method and 35 with the SoA from the initial pool of 45 genes. We perform $5 \times 2$ cross-validation to obtain more stable estimates through repetitions on non-overlapping training sets [37]. We observe an increase in the AUC from $0.60 \pm 0.08$ of the SoA to $0.71 \pm 0.05$ with our attention-based model. In comparison, using the ground-truth sequencing data yields AUC $0.86 \pm 0.04$.

## 4  Discussion

The experiments in Sect. 3.2 quantify the accuracy of our bulk-RNA predictions from WSIs in terms of the prediction error, i.e. MAPE, and correlation with the ground-truth labels, i.e. Pearson correlation $\rho$. Our trainable attention mechanism reduces the error for the majority of the analysed genes, achieving significant improvements over the SoA for the *SPINK4, QPRT, MUC2, SDC2* and *AOC3* genes. Besides accuracy improvements, learning the attention removes the use of a priori heuristic choices and hyperparameter tuning in [11].

The main contribution is the analysis of the signals that are picked by the models to obtain our performance improvements. Focusing on one gene at a time enables a fine-grained analysis of the model attention patterns since the entire model capacity is optimised to capture salient features of individual genes. The attention maps in Sect. 3.3 give interesting insights. For instance, Fig. 3 shows that the attention localises over hotspot regions rather than being uniformly spread across the slide. This answers the questions raised in [17] regarding the distribution of the information related to gene expression. As expected, high attention is seen in overlapping WSI regions for genes that are co-expressed in the same cell and in differing regions for genes that are not co-expressed. Moreover, the highest attention weights point to visibly different histological patterns, as shown in Fig. 5. For example, the high-expression patches of *COL8A2* present higher concentrations of stroma than the low-expression ones.

An interesting example is that of *MUC2*, a particularly well-known gene that impacts tissue histology. *MUC2* over-expression leads to high amounts of visible intra- and extra-cellular mucin, characterizing the tissue as mucinous adenocarcinoma. An example of mucinous adenocarcinoma is shown in Fig. 4b. Our model predicts high values of *MUC2* for the regions in the slides where the mucinous tissue is the most visible. When these are further weighted by the attention, the localisation is highly focused on a few hotspots of high expression. From a quantitative standpoint, the impact of the attention mechanism for this specific gene is remarkably significant. The SoA predictions for *MUC2* are similar to random guessing (with MAPE 0.95 and $\rho$ 0.13), whereas our method significantly reduces the MAPE to 0.73 and the correlation with ground truth values, i.e. $\rho$, to 0.46 with p-value $< 0.0001$. This result is also beneficial to the prediction of MSI, which is known to correlate with the presence of mucin [26].

## 5   Conclusion

We proposed an attention-based multiple instance regression model to infer bulk gene expression of tissue sections from H&E histology slides of colorectal adeno-carcinoma. The automated regression of transcriptomics from H&E is not yet a feasible replacement for sequencing the tissue, but the analysis of interpretability patterns can highlight some histological patterns associated with specific gene expression levels. We show here how the attention mechanism successfully filters informative content from the vast amount of information in the WSIs, leading to significantly lower errors and interesting insights into how gene expression impacts tissue morphology. Further developments of this method should introduce pathologists in the loop to validate the identified patterns and further investigate their association to differentially expressed genes. In turn, this could help pathologists to identify cancer molecular subtypes from WSIs and consequently stratify patients for targeted treatment.

**Acknowledgements.** This work was supported by the Swiss National Science Foundation Sinergia project (CRSII5_193832) and the EU H2020 project AI4Media (951911).

# A    Description of Selected Genes

**Table 1.** Summary of the 45 genes considered in this study, including the ColoType signature and biomarkers of colorectal adenocarcinoma.

| Gene | Expression level | Associated CMS | Ref. | Description |
|------|-----------------|----------------|------|-------------|
| ASPHD2 | HIGH | CMS1 | [22] | Dioxygenase activity |
| ATP9A | LOW | CMS1 | [22] | Membran trafficking of cargo proteins |
| AXIN2 | LOW | CMS1 | [22] | Regulation of beta-catenin stability in WNT pathway |
| CDHR1 | LOW | CMS1 | [22] | Cadherin superfamily, cell adhesion |
| CTTNBP2 | LOW | CMS1 | [22] | Cortacting binding protein |
| DACH1 | LOW | CMS1 | [22] | Chromatin associated protein |
| GNLY | HIGH | CMS1 | [22] | Antimicrobial protein that kills intracellular pathogens |
| HPSE | HIGH | CMS1 | [22] | Enhances angiogenesis |
| SEMA5A | LOW | CMS1 | [22] | Promotes angiogenesis by cell proliferation and migration and inhibites apoptosis |
| WARS | HIGH | CMS1 | [22] | Regulates ERK, Akt, and eNOS pathways, associated with angiogenesis |
| CEL | HIGH | CMS2 | [22] | |
| DDX27 | HIGH | CMS2 | [22] | Probable ATP-dependent RNA helicase |
| DUSP4 | LOW | CMS2 | [22] | Regulates mitogenic signal transduction |
| FSCN1 | LOW | CMS2 | [22] | Organizes filamentous actin into parallel bundles |
| LYZ | LOW | CMS2 | [22] | |
| PLAGL2 | HIGH | CMS2 | [22] | |
| POFUT1 | HIGH | CMS2 | [22] | |
| QPRT | HIGH | CMS2 | [22] | |
| TP53RK | HIGH | CMS2 | [22] | |
| TRIB2 | LOW | CMS2 | [22] | |
| ASRGL1 | HIGH | CMS3 | [22] | |
| B3GNT6 | HIGH | CMS3 | [22] | Plays an important role in the synthesis of mucin-type O-glycans in digestive organs |
| CAPN9 | HIGH | CMS3 | [22] | |
| FBN1 | LOW | CMS3 | [22] | |

(continued)

**Table 1.** (*continued*)

| Gene | Expression level | Associated CMS | Ref. | Description |
| --- | --- | --- | --- | --- |
| FCGBP | HIGH | CMS3 | [22] | |
| RASD1 | HIGH | CMS3 | [22] | |
| RBMS1 | LOW | CMS3 | [22] | |
| SPINK4 | HIGH | CMS3 | [22] | |
| TIMP3 | LOW | CMS3 | [22] | |
| VAV2 | LOW | CMS3 | [22] | |
| AOC3 | HIGH | CMS4 | [22] | Participates in lymphocyte extravasation and recirculation |
| ARMCX1 | HIGH | CMS4 | [22] | Regulates mitochondrial transport during axon regeneration |
| CCDC80 | HIGH | CMS4 | [22] | Promotes cell adhesion and matrix assembly |
| COL8A2 | HIGH | CMS4 | [22] | Necessary for migration and proliferation of vascular smooth muscle cells |
| MGP | HIGH | CMS4 | [22] | Thought to act as inhibitor of bone formation |
| NRP2 | HIGH | CMS4 | [22] | May play a role in cardiovascular development, axon guidance, and tumorigenesis |
| RAB34 | HIGH | CMS4 | [22] | |
| SDC2 | HIGH | CMS4 | [22] | participates in cell proliferation, cell migration and cell-matrix interactions |
| TGFB3 | HIGH | CMS4 | [22] | Involved in embryogenesis and cell differentiation |
| TNS1 | HIGH | CMS4 | [22] | |
| DNMT3B | UNK | UNK | [23] | Required for genome-wide de novo methylation. Seems to be involved in gene silencing |
| CDKN2A | UNK | UNK | [24] | Many studies suggest poorer prognostic outcome for patients with hypermethylation in colorectal, liver, and younger lung cancer patients |
| CEACAM5 | UNK | UNK | [24] | Cell adhesion, intracellular signaling and tumor progression |
| CXCR4 | UNK | UNK | [24] | Essential for the vascularization of the gastrointestinal tract |
| MUC2 | UNK | UNK | [24] | Coats the epithelia in the colon. May exclude bacteria from the inner mucus layer |

# B    Detailed Model Evaluation

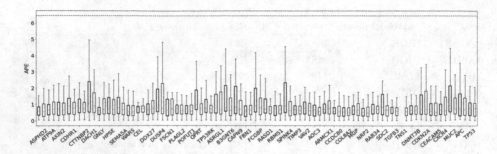

**Fig. 6.** Absolute Percentage Error on the test set for the baseline model (in blue) and our method (in green). The models are trained on five training folds and the final prediction is obtained as the ensembled average of the predictions. The red horizontal line shows the performance of random prediction. The lower the better. (Color figure online)

Detailed results are reported in Table 2, which was not included in the main paper because of the limited space. The model of the gene expression for *NRP2* leads to the best results overall, with MAPE = $0.47 \pm 1.19$ and $\rho = 0.62$ against the SoA MAPE = $0.51 \pm 1.17$ and $\rho = 0.61$ of the baseline.

# C    Additional Visualizations

Additional visualizations of the spatialized prediction maps with and without attention-based weighting are given in Fig. 7. Figures 8 and 9 show in detail the histologic patterns selected by the attention for the genes *COL8A2* and *MUC2* respectively.

# D    Single-Cell Co-expression

The co-expression of genes in single-cell RNA sequencing data is summarized in Fig. 11. Figure 10 compares the correlation of the attention weights in overlapping patches to the co-expression of genes in single-cell data in terms of their Pearson's correlation.

The couples (i) *CCD80, TIMP3*; (ii) *CCDC80, MGP*; (iii) *TIMP3, MGP*; (iv) *FCGBP, MUC2* report high correlation with Pearson's *rho* being 0.85, 0.77, 0.69, 0.62 respectively. These couples show high correlation in the attention patterns. Figure 12 shows the similarity in the attention patterns for *FCGBP* and *MUC2*.

**Table 2.** Comparison of the Test MAPE. Standard deviation reported in brackets.

| Gene | MAPE ↓ | |
|------|--------|------|
|      | SoA | Ours |
| NRP2 | 0.51 (1.2) | 0.47 (1.2) |
| COL8A2 | 0.54 (4.1) | 0.49 (3.5) |
| CEACAM5 | 0.59 (0.8) | 0.52 (0.9) |
| WARS | 0.55 (6.5) | 0.52 (12.3) |
| CCDC80 | 0.56 (2.0) | 0.54 (1.5) |
| TNS1 | 0.54 (1.3) | 0.54 (1.0) |
| ASPHD2 | 0.58 (1.9) | 0.55 (2.7) |
| AOC3 | 0.65 (3.7) | 0.55 (1.3) |
| POFUT1 | 0.70 (0.9) | 0.55 (1.4) |
| TGFB3 | 0.67 (6.2) | 0.57 (2.7) |
| CXCR4 | 0.65 (3.2) | 0.60 (4.1) |
| TIMP3 | 0.66 (3.4) | 0.60 (2.6) |
| VAV2 | 0.74 (1.3) | 0.61 (1.3) |
| FBN1 | 0.58 (1.6) | 0.61 (1.9) |
| PLAGL2 | 0.74 (1.0) | 0.61 (1.4) |
| B3GNT6 | 0.76 (5.0) | 0.62 (2.8) |
| SPINK4 | 0.88 (3.1) | 0.62 (1.9) |
| AXIN2 | 0.71 (0.7) | 0.63 (0.7) |
| DNMT3B | 0.65 (9.3) | 0.64 (10.5) |
| SDC2 | 0.81 (2.3) | 0.64 (1.4) |
| MGP | 0.64 (1.6) | 0.65 (3.0) |
| SEMA5A | 0.80 (1.7) | 0.65 (1.4) |
| RBMS1 | 0.72 (1.1) | 0.65 (1.7) |
| RASD1 | 0.69 (5.8) | 0.66 (4.5) |
| APC | 0.71 (2.1) | 0.67 (2.0) |
| GNLY | 0.62 (4.8) | 0.67 (3.5) |
| ASRGL1 | 0.74 (4.1) | 0.68 (4.1) |
| HPSE | 0.71 (6.7) | 0.68 (5.0) |
| CDKN2A | 0.67 (36.1) | 0.69 (29.8) |
| CAPN9 | 0.82 (5.1) | 0.69 (2.6) |
| RAB34 | 0.69 (2.3) | 0.69 (5.5) |
| ATP9A | 0.64 (0.9) | 0.70 (1.0) |
| FSCN1 | 0.62 (5.0) | 0.70 (4.8) |
| TP53RK | 0.74 (1.2) | 0.71 (0.9) |
| CTTNBP2 | 0.76 (3.0) | 0.71 (2.2) |
| DDX27 | 0.68 (1.0) | 0.72 (2.1) |
| CEL | 0.77 (1.2) | 0.72 (1.4) |
| ARMCX1 | 0.66 (15.1) | 0.73 (15.1) |
| MUC2 | 0.95 (1.9) | 0.73 (1.2) |
| TP53 | 0.76 (0.9) | 0.74 (0.9) |
| DUSP4 | 0.78 (3.1) | 0.74 (3.5) |
| QPRT | 0.98 (1.1) | 0.74 (1.3) |
| CDHR1 | 0.87 (1.2) | 0.76 (1.2) |
| FCGBP | 0.89 (2.5) | 0.77 (2.5) |
| DACH1 | 1.02 (2.9) | 0.90 (2.2) |

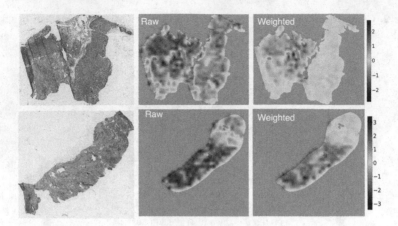

**Fig. 7.** Additional spatialized prediction maps for *AOC3*

# COL8A2

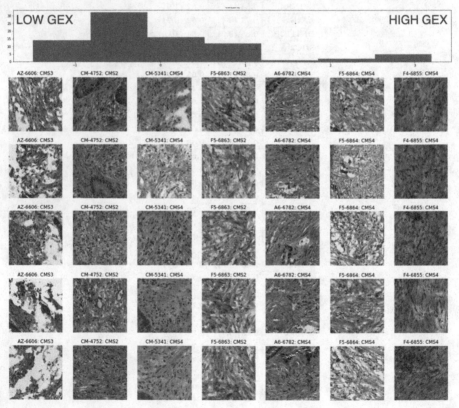

**Fig. 8.** High-attention patches against gene expression levels (gex) for *COL8A2*

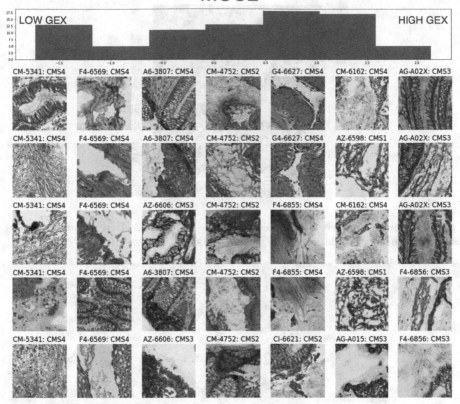

**Fig. 9.** High-attention patches against gene expression levels (gex) for *MUC2*

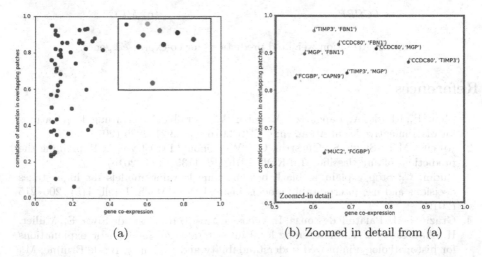

(a)                                              (b) Zoomed in detail from (a)

**Fig. 10.** Correlation of attention patterns in overlapping WSIs regions for co-expressed genes with correlation p-value $< 0.05$.

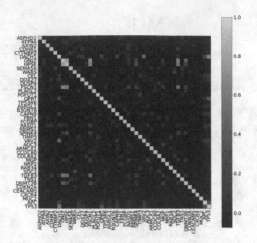

**Fig. 11.** Summary of gene co-expression in single-cells. Each cell in the matrix represents the Pearson's correlation of the gene expressions.

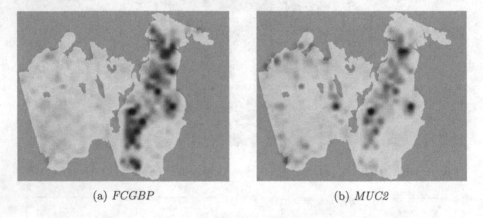

(a) *FCGBP*                    (b) *MUC2*

**Fig. 12.** Attention-weighted predictions for co-expressed genes.

# References

1. Zhou, B., Khosla, A., Lapedriza, A., Oliva, A., Torralba, A.: Learning deep features for discriminative localization. In: CVPR 2016, pp. 2921–2929 (2016)
2. Ribeiro, M.T., Singh, S., Guestrin, C.: "Why should i trust you?": Explaining the predictions of any classifier. In: KDD 2016, pp. 1135–1144 (2016)
3. Rudin, C.: Stop explaining black box machine learning models for high stakes decisions and use interpretable models instead. Nat. Mach. Intell. **1**(5), 206–215 (2019)
4. Graziani, M., Palatnik de Sousa, I., Vellasco, M.M.B.R., Costa da Silva, E., Müller, H., Andrearczyk, V.: Sharpening local interpretable model-agnostic explanations for histopathology: improved understandability and reliability. In: de Bruijne, M., et al. (eds.) MICCAI 2021. LNCS, vol. 12903, pp. 540–549. Springer, Cham (2021). https://doi.org/10.1007/978-3-030-87199-4_51

5. Graziani, M., Andrearczyk, V., Marchand-Maillet, S., Müller, H.: Concept attribution: explaining CNNs to physicians. CBM **123**, 103865 (2020)
6. Hägele, M., et al.: Resolving challenges in deep learning-based analyses of histopathological images using explanation methods. Sci. Rep. **10**(1), 6423 (2020)
7. Graziani, M., Lompech, T., Müller, H., Depeursinge, A., Andrearczyk, V.: On the scale invariance in state of the art CNNs trained on ImageNet. MAKE **3**(2), 374–391 (2021)
8. Graziani, M., Otálora, S., Marchand-Maillet, S., Müller, H., Andrearczyk, V.: Learning interpretable pathology features by multi-task and adversarial training improves CNN generalization. arXiv:2008.01478 (2021)
9. McGrath, T., et al.: Acquisition of chess knowledge in alphazero. ArXiv:2111.09259 (2021)
10. Chen, M., et al.: Classification and mutation prediction based on histopathology H&E images in liver cancer using deep learning. NPJ Precis. Oncol. **4**(1), 1–7 (2020)
11. Schmauch, B., et al.: A deep learning model to predict RNA-Seq expression of tumours from whole slide images. Nat. Commun. **11**(1), 1–15 (2020)
12. van der Velden, B.H., Kuijf, H.J., Gilhuijs, K.G., Viergever, M.A.: Explainable artificial intelligence (XAI) in deep learning-based medical image analysis. MIA **79**, 102470 (2022)
13. Lengerich, B.J., Caruana, R., Nunnally, M.E., Kellis, M.: Death by round numbers and sharp thresholds: how to avoid dangerous AI EHR recommendations. medRxiv (2022)
14. Guinney, J., et al.: The consensus molecular subtypes of colorectal cancer. Nat. Med. **21**(11), 1350–1356 (2015)
15. Trullas, A., et al.: The EMA assessment of pembrolizumab as monotherapy for the first-line treatment of adult patients with metastatic microsatellite instability-high or mismatch repair deficient colorectal cancer. ESMO Open **6**(3), 100145 (2021)
16. Ilse, M., Tomczak, J., Welling, M.: Attention-based deep multiple instance learning. In: ICML, pp. 2127–2136. PMLR (2018)
17. Weitz, P., Wang, Y., Hartman, J., Rantalainen, M.: An investigation of attention mechanisms in histopathology whole-slide-image analysis for regression objectives. In: ICCV, pp. 611–619 (2021)
18. Cooper, L.A.D., Demicco, E.G., Saltz, J.H., Powell, R.T., Rao, A., Lazar, A.J.: Pancancer insights from the cancer genome atlas: the pathologist's perspective. J. Pathol. **244**(5), 512–524 (2018)
19. Janowczyk, A., Zuo, R., Gilmore, H., Feldman, M., Madabhushi, A.: HistoQC: an open-source quality control tool for digital pathology slides. JCO Clin. Cancer Inform. **3**, 1–7 (2019)
20. Goldman, M.J., et al.: Visualizing and interpreting cancer genomics data via the Xena platform. Nat. Biotechnol. **38**(6), 675–678 (2020)
21. Grossman, R.L., et al.: Toward a shared vision for cancer genomic data. N. Engl. J. Med. **375**(12), 1109–1112 (2016)
22. Buechler, S.A., et al.: ColoType: a forty gene signature for consensus molecular subtyping of colorectal cancer tumors using whole-genome assay or targeted RNA-sequencing. Sci. Rep. **10**(1), 1–13 (2020)
23. Pan, F., et al.: Prognosis prediction of colorectal cancer using gene expression profiles. Front. Oncol. **79**, 2414 (2019)
24. Kheirelseid, E.A.H., Miller, N., Chang, K.H., Nugent, M., Kerin, M.J.: Clinical applications of gene expression in colorectal cancer. J. Gastrointest. Oncol. **4**(2), 144 (2013)

25. Li, H., et al.: Reference component analysis of single-cell transcriptomes elucidates cellular heterogeneity in human colorectal tumors. Nat. Genet. **49**(5), 708–718 (2017)
26. Nguyen, H.G., et al.: Image-based assessment of extracellular mucin-to-tumor area predicts consensus molecular subtypes (CMS) in colorectal cancer. Mod. Pathol. **35**(2), 240–248 (2022)
27. Marini, N., et al.: Unleashing the potential of digital pathology data by training computer-aided diagnosis models without human annotations. npj Digit. Med. **5**(1), 1–18 (2022)
28. Marini, N., et al.: Multi_Scale_Tools: a python library to exploit multi-scale WSIs. Data-Enabled Intelligence for Medical Technology Innovation, VI (2022)
29. Soumya, R., David, P.: Multiple instance regression. In: ICML, pp. 425–432. Morgan Kaufmann (2001)
30. He, K., Zhang, X., Ren, S., Sun, J.: Deep residual learning for image recognition. In: CVPR, pp. 770–778 (2016)
31. Russakovsky, O., et al.: ImageNet large scale visual recognition challenge. IJCV **115**(3), 211–252 (2015). https://doi.org/10.1007/s11263-015-0816-y
32. Schlemper, J., et al.: Attention gated networks: learning to leverage salient regions in medical images. MIA **53**, 197–207 (2019)
33. Jain, S., Wallace, B.C.: Attention is not explanation. In: Proceedings of NAACL-HLT, pp. 3543–3556 (2019)
34. Wiegreffe, S., Pinter, Y.: Attention is not not explanation. In: EMNLP-IJCNLP, Hong Kong, China, pp. 11–20. Association for Computational Linguistics, November 2019
35. Haab, J., Deutschmann, N., Martínez, M.R.: Is attention interpretation? A quantitative assessment on sets. arXiv preprint arXiv:2207.13018 (2022)
36. Lacoste, A., Luccioni, A., Schmidt, V., Dandres, T.: Quantifying the carbon emissions of machine learning. arXiv:1910.09700 (2019)
37. Alpaydmn, E.: Combined $5 \times 2$ CV F test for comparing supervised classification learning algorithms. Neural Comput. **11**(8), 1885–1892 (1999)

# Beyond Voxel Prediction Uncertainty: Identifying Brain Lesions You Can Trust

Benjamin Lambert[1,2(✉)], Florence Forbes[3], Senan Doyle[2], Alan Tucholka[2], and Michel Dojat[1]

[1] Univ. Grenoble Alpes, Inserm, U1216, Grenoble Institut Neurosciences, 38000 Grenoble, France
benjamin.lambert@univ-grenoble-alpes.fr
[2] Pixyl, Research and Development Laboratory, 38000 Grenoble, France
[3] Univ. Grenoble Alpes, Inria, CNRS, Grenoble INP, LJK, 38000 Grenoble, France

**Abstract.** Deep neural networks have become the gold-standard approach for the automated segmentation of 3D medical images. Their full acceptance by clinicians remains however hampered by the lack of intelligible uncertainty assessment of the provided results. Most approaches to quantify their uncertainty, such as the popular Monte Carlo dropout, restrict to some measure of uncertainty in prediction at the voxel level. In addition not to be clearly related to genuine medical uncertainty, this is not clinically satisfying as most objects of interest (e.g. brain lesions) are made of groups of voxels whose overall relevance may not simply reduce to the sum or mean of their individual uncertainties. In this work, we propose to go beyond voxel-wise assessment using an innovative Graph Neural Network approach, trained from the outputs of a Monte Carlo dropout model. This network allows the fusion of three estimators of voxel uncertainty: entropy, variance, and model's confidence; and can be applied to any lesion, regardless of its shape or size. We demonstrate the superiority of our approach for uncertainty estimate on a task of Multiple Sclerosis lesions segmentation.

**Keywords:** MS lesion · Detection · Deep learning · Interpretabilty · Prediction

## 1 Introduction

Magnetic Resonance Imaging (MRI) is the standard imaging modality for the diagnosis and follow-up of Multiple Sclerosis (MS). It allows a direct observation of brain lesions produced by the disease and provides information about the pathology stage or treatment efficiency. Deep Learning (DL) approaches, based on a trained U-Net-like neural network, are invaluable tools to automatically delineate MS lesions [18]. Although powerful and versatile, these models provide segmentation maps that are typically opaque, with no indication regarding

M. Reyes et al. (Eds.): iMIMIC 2022, LNCS 13611, pp. 61–70, 2022.
https://doi.org/10.1007/978-3-031-17976-1_6

their certainty. This hinders full acceptance of DL models in clinical routine, for which uncertainty attached to the computerized results is essential for their interpretation and to avoid misleading predictions.

A variety of methods have been proposed to quantify the uncertainty attached to deep neural networks [1]. Among them, the Monte Carlo (MC) dropout stands out as one of the simplest approach, as it can be applied to any model trained with the dropout technique [20]. Such a model can be interpreted as a Bayesian neural network, giving access to the interesting properties of these probabilistic models regarding quantification of their uncertainty [4]. More particularly at inference, for a given input, multiple stochastic forward passes are computed by keeping dropout activated, corresponding to empirical samples from the approximated predictive distribution. This produces a set of softmax probabilities that can further be used to compute uncertainty estimates. Applied to MRI segmentation, the MC dropout method produces uncertainty metrics for each voxel in the volume, resulting in so-called voxel-wise uncertainty maps [6,12,17]. The clinically-relevant information, however, is at a higher level, typically at the instance (lesion, tissue) level.

Natural ways to obtain such instance-wise uncertainties, meaning the uncertainties attached to each connected component within the output segmentation, are through a *post hoc* aggregation of voxel-wise uncertainty estimations. Existing approaches include computing the mean uncertainty of voxels belonging to the same class in the segmentation [16] (thus producing one uncertainty estimate per class, rather than per connected component). In the context of MS, lesion-wise uncertainty was also estimated using the logsum of the connected voxels uncertainties [12]. Using the mean implies that each component uncertainty contributes equally to the overall instance score, while the use of the logsum assumes that connected voxels are conditionally independent, given that they belong to the same instance. These highly simplified assumptions may degrade the quality of instance uncertainty computation. To go further, a side-learner called MetaSeg has been proposed to predict the Intersection Over Unions (IoU) of each individual segmented instance with the ground truth [15]. For this task, a Linear Regression Model is trained based on a series of features derived from a standard segmentation model's output probabilities. The predicted score is then used as a marker of instance uncertainty. Yet, the input features of MetaSeg consist in averaged voxel-wise metrics, leading to the same restrictions than the previously-described *post hoc* aggregation methods. Additionally, it has been proposed to train an auxiliary Graph (Convolutional) Neural Network (GCNN) using the outputs of a MC dropout U-Net (i.e. voxel-wise segmentation and uncertainty maps) to refine the predicted masks [19]. This approach, however, remains at the voxel level and focuses on 2D segmentation tasks.

In this work, we propose to build from the two last methods to overcome their respective limitations. Indeed, we implement a GCNN at the output of a trained MC dropout U-Net model. Using the predicted 3D segmentation outputs, each individual segmented lesion is modeled by a graph whose voxels are the interconnected nodes. Node features are determined by the input and output of the U-Net, comprising the voxel image intensities, the voxel predicted label, and voxel-wise uncertainty maps. We implement two alternative variants of the proposed GCNN, either classification or regression, to quantify lesions uncertainty. We test our framework on a task of 3D binary segmentation on MS data. Results demonstrate the superiority of our approach compared to known methods.

## 2 Our Framework: Graph Modelization for Lesion Uncertainty Quantification

*Overview:* Consider an input image X and a trained MC dropout segmentation model $\mathcal{N}$ with parameters $W$ that produces a segmentation $Y = \mathcal{N}(X, W)$ and a set of voxel-wise uncertainty maps $U_i$ (e.g. entropy, variance, PCS, etc.). Our objective is to quantify the uncertainty of each instance (i.e. lesion) in Y. To do so, we propose to train an auxiliary GCNN to predict this uncertainty directly from X, Y, and $U_i$ (see Fig. 1).

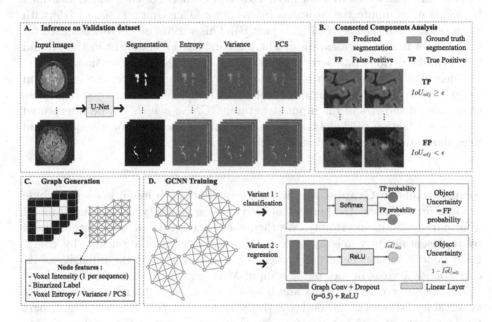

**Fig. 1.** Illustration of the proposed framework for learning lesion uncertainty from the outputs of a Monte Carlo dropout model. See the text for details of each block.

### 2.1 Monte Carlo Dropout Model and Voxel-Wise Uncertainty

We use a generic 3D U-Net [2] for its simplicity and popularity within the field, although our method can be employed with any segmentation model trained with dropout. We add 3D dropout [21] with a rate of $p = 0.2$ at the end of each encoding and decoding block. The model is trained on annotated datasets composed of pairs of images: (i) input T2-weighted FLAIR MRI sequences X and

(ii) associated ground truth MS lesions segmentation $Y$. At inference, dropout is kept activated and $T$ forward passes are made for a new input volume $x^*$. We chose $T = 20$, as it allows an optimal counterpart between inference time and quality of uncertainty estimates [13]. From this set of predictions, several well-known voxel-related uncertainty metrics are extracted: (see Fig. 1, part A): the entropy [5], the variance [7] and the Predicted Confidence Score (PCS) [23].

## 2.2    Graph Dataset Generation

**Inference on Validation Dataset and Connected Component Analysis.**
After training, the MC dropout U-Net is subsequently used to generate segmentation and uncertainty maps on the set-aside validation set of images. These predictions are used to generate training data for the auxiliary GCNN. We use a Connected Component Analysis (CCA) to identify each lesion in the segmentation masks using 26-connectivity—meaning that a lesion is defined by voxels that are interconnected by their faces, edges, or corner. For each lesion identified by CCA, we compute the Adjusted Intersection Over Union ($IoU_{adj}$) [15] with the ground truth lesions (see Fig. 1, part B). This variant of the IoU is suited for brain-abnormalities segmentation, where a connected component in the ground truth can be divided into several pieces in the predicted segmentation.

Identified lesions can exhibit a wide range of shape and size. To learn from these data, we must thus design a neural network that can be employed regardless of the shape and size of the input structure. GCNNs, which can be interpreted as a generalization of the classic convolutional networks to non-Euclidean and irregular data, are thus particularly suitable for this task.

**From Voxels to Graphs.** We first slightly dilate each lesion mask to include surrounding voxels at the border between classes, which typically convey useful information about uncertainty. We then convert the dilated mask into a graph by representing its voxels by nodes and neighborhood relationships by edges. Each node is further defined by a set of $n + 4$ features: (i) the intensity of its corresponding voxel in each of the $n$ input MRI sequences, (ii) its binarized label (1 for the observed lesion class and 0 for all other classes), and its 3 voxel-wise uncertainty estimates: (iii) entropy, (iv) variance and (v) PCS (see Fig. 1, part C). In agreement with the aforementioned 26-connectivity CCA, each node (i.e. voxel) is connected in the graph to its 26 nearest neighbors.

## 2.3    GCNN Architecture and Training

Here, we use a lightweight GCNN architecture composed of 2 consecutive Graph Convolutional layers with a hidden dimension of $h = 64$, followed by a Linear layer (see Fig. 1, part D). The model is trained using the graph dataset generated from the validation images, composed of graphs (transformed connected components obtained from the segmentation model) along with their associated ground truth ($IoU_{adj}$). As in [15], we propose two versions of our model:

– In the classification approach (GCNN$_{\text{Classif}}$), the $IoU_{adj}$ labels are first binarized as follows: FP if $IoU_{adj}(graph) < \epsilon$, and TP if $IoU_{adj}(graph) \geq \epsilon$. $\epsilon$ is a
  hyperparameter that we set to 0.1 in our experiments, so that lesions with an
  $IoU_{adj}$ very close to 0 are not wrongly considered as TP. The network is then
  trained using the Cross-Entropy Loss. At inference, structural uncertainty is
  quantified by the graph FP probability.
– In the regression approach (GCNN$_{\text{Reg}}$), the model is directly trained to predict the graph $\widehat{IoU}_{adj}$, using the MSE loss. At inference, we use $1 - \widehat{IoU}_{adj}$
  as the structural uncertainty score.

## 3   Material and Method

### 3.1   Data

We combine two open-source MS datasets: from the University Hospital of Ljubl-
jana (MSLUB) [10] and from the MICCAI 2016 MS segmentation challenge
(MSSEG 2016) [3]. We thus use 83 manually-annotated 3D T2-FLAIR sequences.
Images are resampled to a 1 mm isotropic resolution of $160 \times 192 \times 160$ to focus
on brain tissues, and intensities are normalized to zero mean and unit variance.
We opt for a 4-fold cross-validation scheme due to the limited number of images.
In each fold, we put aside 25% of the images for testing. From the remaining
images, we use 20% for validation and 80% to train the model. During evaluation, results are averaged over the 4 folds. Due to the limited number of images,
we extensively use Data Augmentation to train our models, comprising flipping,
rotation, contrast alteration, gaussian noise and blurring.

### 3.2   Comparison with Known Approaches

To evaluate the relevance of our proposed GCNN$_{\text{Classif}}$ and GCNN$_{\text{Reg}}$
approaches, we implement in parallel known approaches to obtain instance uncertainty from the U-Net. We use the mean and logsum of the voxel-wise uncertainty of each lesions, with the 3 different types of uncertainty. We name these
methods Entropy$_{\text{mean}}$, Variance$_{\text{mean}}$, PCS$_{\text{mean}}$, Entropy$_{\text{logsum}}$, Variance$_{\text{logsum}}$,
and PCS$_{\text{logsum}}$.

   As pointed out in [12], using the logsum assigns a higher uncertainty to small-
size lesions. This appears sub-optimal as small lesions could be segmented with
high confidence, especially in the case of MS lesions. To verify this point, we
implement a naive approach, named Size, which attributes a lesion uncertainty
inversely proportional to its size. The lesion size (number of voxels composing
it) being $S$, its uncertainty is computed as $1/S$.

   Lastly, we implement an approach inspired from the MetaSeg framework [15].
We extract a series of features from each connected component in the validation
dataset, consisting in the mean entropy, variance and PCS, as well as the size
of the lesion. We then train a Logistic Regression classifier from these 4 features to distinguish between True Positive (TP) and FP lesions (MetaSeg$_{\text{Classif}}$).

Alternatively, we train a Linear Regression model to directly predict $\widehat{IoU}_{adj}$ (MetaSeg$_{Reg}$). We use the outputs of these models to obtain lesion uncertainty as described for the GCNN approach.

### 3.3 Evaluation Setting

For medical applications, the ideal uncertainty quantification should attribute a higher uncertainty to FP lesions than TP, to allow for proper interpretation and evaluation of the results. To evaluate this properly, we use Accuracy-Confidence curves [9]. Briefly, the principle is to set aside the $\tau\%$ of the most uncertain predicted lesions among the test dataset, and measure the performance of the model on the remaining lesions by counting the number of FP and TP lesions. The threshold $\tau$ fluctuates between 0 (all lesions are kept) and 100 (all lesions are removed). By plotting the couples (FP, TP) at different thresholds, we obtain an Accuracy-Confidence curve and compute the AUC (Area Under the Curve) score reflecting the quality of the estimated lesion uncertainty. FP and TP counts are normalized in the range $[0, 1]$ by dividing by the counts obtained without filtering (at $\tau = 0$). This metric only depends on the ranking of uncertainties, thus is independent of the uncertainty ranges of each method ensuring a fair comparison. We additionally evaluate the segmentation performance of the U-Net on the test datasets using the Dice coefficient, as well as the total number of TP and FP lesions. Finally, for each method, we control the correlation between the estimated uncertainty and the lesion size using the Spearman's rank correlation coefficient $(\rho)$.

**Table 1.** U-Net segmentation performance on the MS dataset and number of TP and FP lesions for each fold.

| Fold | 0 | 1 | 2 | 3 |
|---|---|---|---|---|
| Dice | 0.672 | 0.645 | 0.705 | 0.693 |
| # TP lesions | 829 | 597 | 715 | 871 |
| # FP lesions | 525 | 294 | 353 | 454 |

### 3.4 Implementation Details

**3D Segmentation U-Net.** Our segmentation framework was implemented using PyTorch [14]. We opt for a patch approach to train the segmentation U-Net, meaning that the $160 \times 192 \times 160$ MRI volumes are split into 3d patches of $160 \times 192 \times 32$, decreasing the memory cost of training. We use a batch size of 5. The U-Net is trained with a combination of the Dice [11] and Cross-Entropy loss, using the ADAM optimizer [8] with a learning rate of $1e^{-4}$ until convergence. For the training of the segmentation models, a single NVIDIA T4 GPU was used.

**Graph Neural Networks.** We use the Deep Graph Library [22] to implement and train the GCNN models. The training procedure of our GCNN is standard: we use the ADAM optimizer with a learning rate of $1e^{-2}$ at the start of training, and progressively decreasing to $1e^{-5}$. Graphs are presented to the network in the form of batches, composed of 10 graphs. Due to the small size of the GCNN models, they were trained on CPU, which took a couple of minutes in our experiments.

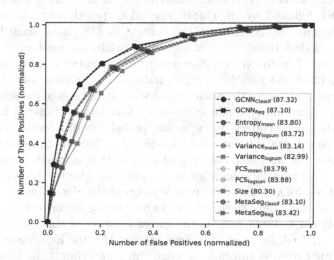

**Fig. 2.** Accuracy-Confidence curves for the different methods. The associated AUC scores are indicated in brackets in the graphs legends.

## 4    Results and Discussion

Accuracy-Confidence curves are presented in Fig. 2 along with the corresponding AUC values. Segmentation performance and correlation coefficients are presented in Tables 1 and 2. Experimental results show that both models of the proposed framework outperform the classical methods by a significant margin, and that their performances are similar with a very small advantage for the classification version. The naive Size approach achieves the lowest performance. Similarly, the *logsum* approaches, also strongly correlated with the lesion size, have poorer performance than the *mean* counterparts. Not surprisingly, in the context of MS lesions, the lesion size is not a satisfying proxy for uncertainty as small lesions can be segmented with high confidence. In our experimental setting, *MetaSeg* models do not outperform simpler methods. This is probably due to the overall simplicity of these models, failing to fully learn the relationships between the different input features.

Results show that our graph-based framework can be efficiently used to flag uncertain lesions, that are more likely to result in False Positives. The classification variant slightly outperforms regression. We hypothesize that this is due to the increased difficulty of predicting the exact $IoU_{adj}$, compared to the binary classification setting. One drawback of our approach is that it requires an additional validation set containing enough lesions (typically a few hundreds) to allow GCNN training. However, as most DL pipelines rely on a set-aside validation set to control overfitting during training, these data can then be used for this purpose (as it was the case in this work). The requirement is thus not prohibitive and only necessitates a sufficiently large validation set.

Overall, our framework is computationally light as CCA is computed only once per MRI, followed by the graph generation step that can be parallelized among the lesions. Additionally, in the context of MS, most brain lesions are relatively small (less than 1000 voxels), which results in small graphs that are fast to generate. Finally we use 26-connectivity, meaning that a voxel is only connected to its closest neighbors, which reduces the computational burden.

Our approach enhances the binary voxel-wise predictions of the segmentation model with reliable and readable lesion-wise uncertainty estimates. In the classification setting, uncertainty is cast as the probability of a lesion being a false positive, which is a straightforward and intelligible definition. In a real world clinical application, this may help the clinician examine the automated segmentation in the light of the model's confidence, hence allowing a better interpretability of the provided results and a more trustable usage of the algorithm.

Future work will study the extension to multi-class segmentation, and inclusion of additional features such as the global location of the lesion within the MRI volume. Indeed, for brain disorders such as MS, the location of the lesion within the brain conveys information concerning uncertainty, as false positives tend to be concentrated in specific brain regions.

**Table 2.** Evaluation of uncertainty estimates (AUC values). $\rho$ represents Spearman's rank correlation coefficient $\rho$.

|  | AUC (%) | Spearman's $\rho$ |
|---|---|---|
| GCNN$_{Classif}$ | **87.32** | −0.78 |
| GCNN$_{Reg}$ | 87.10 | −0.77 |
| Entropy$_{mean}$ | 83.80 | −0.42 |
| Entropy$_{logsum}$ | 83.72 | −0.97 |
| Variance$_{mean}$ | 83.14 | −0.44 |
| Variance$_{logsum}$ | 82.99 | −0.99 |
| PCS$_{mean}$ | 83.79 | −0.44 |
| PCS$_{logsum}$ | 83.88 | −0.98 |
| Size | 80.30 | −1.00 |
| MetaSeg$_{Classif}$ | 83.10 | −0.76 |
| MetaSeg$_{Reg}$ | 83.42 | −0.77 |

# 5  Conclusion

This paper presents an innovative graph-based framework to quantify lesion-wise uncertainty. We demonstrate, with our approach, improvement of the predicted uncertainty, when compared to various known methods. The strength of our solution is its generic nature, making it compatible with any segmentation model trained with dropout.

# References

1. Abdar, M., et al.: A review of uncertainty quantification in deep learning: techniques, applications and challenges. Inf. Fusion **76**, 243–297 (2021)
2. Çiçek, Ö., Abdulkadir, A., Lienkamp, S.S., Brox, T., Ronneberger, O.: 3D U-Net: learning dense volumetric segmentation from sparse annotation. In: Ourselin, S., Joskowicz, L., Sabuncu, M.R., Unal, G., Wells, W. (eds.) MICCAI 2016. LNCS, vol. 9901, pp. 424–432. Springer, Cham (2016). https://doi.org/10.1007/978-3-319-46723-8_49
3. Commowick, O., et al.: Multiple sclerosis lesions segmentation from multiple experts: the MICCAI 2016 challenge dataset. Neuroimage **244**, 118589 (2021)
4. Gal, Y., Ghahramani, Z.: Dropout as a Bayesian approximation: Representing model uncertainty in deep learning. In: International Conference on Machine Learning, pp. 1050–1059. PMLR (2016)
5. Gal, Y., Islam, R., Ghahramani, Z.: Deep Bayesian active learning with image data. In: International Conference on Machine Learning, pp. 1183–1192. PMLR (2017)
6. Jungo, A., et al.: Towards uncertainty-assisted brain tumor segmentation and survival prediction. In: Crimi, A., Bakas, S., Kuijf, H., Menze, B., Reyes, M. (eds.) BrainLes 2017. LNCS, vol. 10670, pp. 474–485. Springer, Cham (2018). https://doi.org/10.1007/978-3-319-75238-9_40
7. Kendall, A., Badrinarayanan, V., Cipolla, R.: Bayesian SegNet: model uncertainty in deep convolutional encoder-decoder architectures for scene understanding. In: British Machine Vision Conference 2017, BMVC 2017, London, UK, 4–7 September 2017. BMVA Press (2017)
8. Kingma, D.P., Ba, J.: Adam: a method for stochastic optimization. In: Bengio, Y., LeCun, Y. (eds.) 3rd International Conference on Learning Representations, ICLR 2015, San Diego, CA, USA, 7–9 May 2015, Conference Track Proceedings (2015). http://arxiv.org/abs/1412.6980
9. Lakshminarayanan, B., Pritzel, A., Blundell, C.: Simple and scalable predictive uncertainty estimation using deep ensembles. In: Advances in Neural Information Processing Systems, vol. 30, Annual Conference on Neural Information Processing Systems 2017, pp. 6402–6413 (2017)
10. Lesjak, Ž., et al.: A novel public MR image dataset of multiple sclerosis patients with lesion segmentations based on multi-rater consensus. Neuroinformatics **16**(1), 51–63 (2018)
11. Milletari, F., Navab, N., Ahmadi, S.A.: V-net: fully convolutional neural networks for volumetric medical image segmentation. In: 2016 Fourth International Conference on 3D Vision (3DV), pp. 565–571. IEEE (2016)

12. Nair, T., Precup, D., Arnold, D.L., Arbel, T.: Exploring uncertainty measures in deep networks for multiple sclerosis lesion detection and segmentation. In: Frangi, A.F., Schnabel, J.A., Davatzikos, C., Alberola-López, C., Fichtinger, G. (eds.) MICCAI 2018. LNCS, vol. 11070, pp. 655–663. Springer, Cham (2018). https://doi.org/10.1007/978-3-030-00928-1_74
13. Orlando, J.I., et al.: U2-net: a Bayesian u-net model with epistemic uncertainty feedback for photoreceptor layer segmentation in pathological oct scans. In: 2019 IEEE 16th International Symposium on Biomedical Imaging (ISBI 2019), pp. 1441–1445. IEEE (2019)
14. Paszke, A., et al.: Pytorch: an imperative style, high-performance deep learning library. In: Advances in Neural Information Processing Systems, vol. 32, pp. 8024–8035. Curran Associates, Inc. (2019)
15. Rottmann, M., et al.: Prediction error meta classification in semantic segmentation: Detection via aggregated dispersion measures of softmax probabilities. In: 2020 International Joint Conference on Neural Networks (IJCNN), pp. 1–9. IEEE (2020)
16. Roy, A.G., et al.: Bayesian QuickNAT: model uncertainty in deep whole-brain segmentation for structure-wise quality control. Neuroimage **195**, 11–22 (2019)
17. Sander, J., de Vos, B.D., Wolterink, J.M., Išgum, I.: Towards increased trustworthiness of deep learning segmentation methods on cardiac MRI. In: Medical Imaging 2019: Image Processing, vol. 10949, p. 1094919. International Society for Optics and Photonics (2019)
18. Shoeibi, A., et al.: Applications of deep learning techniques for automated multiple sclerosis detection using magnetic resonance imaging: a review. Comput. Biol. Med. **136**, 104697 (2021)
19. Soberanis-Mukul, R.D., Navab, N., Albarqouni, S.: Uncertainty-based graph convolutional networks for organ segmentation refinement. In: Medical Imaging with Deep Learning, pp. 755–769. PMLR (2020)
20. Srivastava, N., Hinton, G., Krizhevsky, A., Sutskever, I., Salakhutdinov, R.: Dropout: a simple way to prevent neural networks from overfitting. J. Mach. Learn. Res. **15**(1), 1929–1958 (2014)
21. Tompson, J., Goroshin, R., Jain, A., LeCun, Y., Bregler, C.: Efficient object localization using convolutional networks. In: Proceedings of the IEEE Conference on Computer Vision and Pattern Recognition, pp. 648–656 (2015)
22. Wang, M., et al.: Deep graph library: a graph-centric, highly-performant package for graph neural networks. arXiv preprint arXiv:1909.01315 (2019)
23. Zhang, X., et al.: Towards characterizing adversarial defects of deep learning software from the lens of uncertainty. In: 2020 IEEE/ACM 42nd International Conference on Software Engineering (ICSE), pp. 739–751. IEEE (2020)

# Interpretable Vertebral Fracture Diagnosis

Paul Engstler[1], Matthias Keicher[1], David Schinz[2], Kristina Mach[1],
Alexandra S. Gersing[2,3], Sarah C. Foreman[2], Sophia S. Goller[3],
Juergen Weissinger[3], Jon Rischewski[3], Anna-Sophia Dietrich[2],
Benedikt Wiestler[2], Jan S. Kirschke[2], Ashkan Khakzar[1(✉)],
and Nassir Navab[1,4]

[1] Technical University of Munich, Munich, Germany
ashkan.khakzar@tum.de
[2] Klinikum Rechts der Isar (Technical University of Munich), Munich, Germany
[3] Klinikum der Universität München (University of Munich), Munich, Germany
[4] Johns Hopkins University, Baltimore, USA

**Abstract.** Do black-box neural network models learn clinically relevant
features for fracture diagnosis? The answer not only establishes reliability, quenches scientific curiosity, but also leads to explainable and verbose findings that can assist the radiologists in the final and increase
trust. This work identifies the concepts networks use for vertebral fracture diagnosis in CT images. This is achieved by associating concepts to
neurons highly correlated with a specific diagnosis in the dataset. The
concepts are either associated with neurons by radiologists pre-hoc or
are visualized during a specific prediction and left for the user's interpretation. We evaluate which concepts lead to correct diagnosis and which
concepts lead to false positives. The proposed frameworks and analysis
pave the way for reliable and explainable vertebral fracture diagnosis.
The code is publicly available (https://github.com/CAMP-eXplain-AI/
Interpretable-Vertebral-Fracture-Diagnosis).

**Keywords:** Vertebral fracture diagnosis · Interpretability

## 1 Introduction

Osteoporosis is regarded as one of the most relevant diseases of the elderly, with
22 million women and 5.5 million men affected in the EU alone [5,14]. Early
detection of incidental osteoporotic fractures in routinely-acquired computed
tomography (CT) scans is important, as these often remain clinically silent for a
long time [12]. Furthermore, osteoporotic fractures are an independent predictor
of further fractures with an approx. 12-fold increased risk and are associated with
an 8-fold increased mortality [6,24]. The sequelae include major socioeconomic
consequences and an individual reduction in quality of life [4,7,13,16]. Despite

P. Engstler and M. Keicher—Both authors share first authorship.

the clinical significance, around 85% of osteoporotic fractures are not adequately described in the radiological reports of routinely acquired CT scans, possibly as a result of a disproportionate increase in radiologists' workload [2,31].

Automatic detection of vertebral body fractures with deep learning models can remedy this and increase incidental findings. However, most of these methods are black-box models that do not give insights into the decision-making process. Revealing the inside of these models can allow for investigation of failure cases and, when addressed, increase robustness and trust in the system.

Thus far, interpretable diagnosis is mostly investigated via feature attribution (saliency) approaches [19] such as class activation maps [36]. These interpretations reveal where important features for the prediction are located. Although being a valuable tool for running a sanity check on the network inference mechanism, feature attribution does not disclose further information regarding prediction. Moreover, only knowing about the location of important features is not useful information for fracture diagnosis as it is easy to see where the fracture is located, and it is of interest to know "what" features are important.

To this end, we leverage the network dissection [3] approach and analyze the internal units of the neural network and their associated clinical concepts, inspired by its applications in chest radiography [19] and mammography [32]. Subsequently, we ask the clinicians to identify the concepts associated with highly correlated activations by inspecting the inputs that activate those neurons the highest. We investigate what concepts the network has learned and whether they are aligned with what clinicians use. Moreover, we visualize the concepts used for prediction on a single input to get a conceptual understanding of the decision-making mechanism of the model. We perform the analysis for on the open-source VerSe [29] dataset and a larger private dataset procured at our hospitals. The objective of this work is to investigate what features the network uses for fracture diagnosis, whether they overlap with clinical knowledge, and how they can be used for more verbose and explainable fracture diagnosis.

### 1.1   Related Work

**Vertebral Fracture Detection.** Most approaches use Convolutional Neural Networks (CNN) on Computer Tomography (CT) spine images. CNN-based methods can be categorized into 2D and 3D convolutions. 2D methods usually rely on a feature aggregation with Recurrent Neural Networks to model inter-slice dependencies [1,30]. Husseini et al. [15] reformat the image to use the most informative mid-sagittal slice of each vertebra and, in addition to fracture detection, grade fractures using an ordinal regression loss for representation learning. Pisov et al. [27] also reformat the 3D volume to retrieve a spine-centered 2D image and detect key points for measuring the compression of each vertebra, detecting and grading fractures.

Detecting fractures on a voxel-level and then post-processing, Nicolaes et al. [26] for the first time used 3D convolutions for the detection of vertebral fractures. More recent works using 3D convolutions include modeling the dependency between the 3D volumes of each vertebra with a sequence-to-sequence model [8] and detecting osteoporotic fractures on a patient-level [33]. Related to the task of fracture detection and grading, recently Li et al. [22], and Feng et al. [10] explored the distinction between benign and malign vertebral fractures.

**Interpretability** of models is narrowly explored in the domain of vertebral fracture diagnosis and [34] interprets the models by feature attribution (saliency) approaches to identify which regions in the input contributed to the prediction. In fact, in most medical image analysis applications, feature attribution is the dominant approach [19]. However, attribution methods are limited in the information they can disclose regarding the decision-making mechanism of the model. Moreover, the feature attribution problem remains largely unsolved, and although there are many attribution approaches (CAM [36], LRP [25], DeepSHAP [23], IBA [20,28,35]...), the methods disagree with the identified important features [17,18,35]. This disagreement problem is a caveat for domain experts while utilizing these attribution methods. Thus there is a need for interpretation approaches that are reliable and reveal more information than "which region is important." Network Dissection [3,19,32] allows to identify the concepts encoded by internal units (neurons) of the network. Methodologically, our work differs from [3,19] in that we do not use an annotation dataset and instead identify the highly correlated neurons with the output. Furthermore, the main contribution of this work is establishing trust by investigating the alignment between the learned concepts and vertebrae fracture analysis domain knowledge.

## 2 Methodology

### 2.1 Vertebral Fracture Detection

We model the vertebral fracture detection task as a binary classification problem, where the positive class indicates a fracture. The network function is defined as $f_\Theta(x) : \mathbb{R}^{H \times W \times D} \to \mathbb{R}$. The predicted probability is $\hat{y} = sigmoid(f_\Theta(x))$. We use a 3D U-Net [9] for the vertebral fracture classification task, replacing its upsampling path with a classification head.

### 2.2 Semantic Concept Extraction (Correlation)

In neural networks, each neuron is activated by a specific input pattern. The corresponding pattern of each neuron can be equivalently deemed as its associated *concept*. In convolutional neural networks, each neuron can be considered either as an activation map or an activation unit within the map. As the activation units within an activation map all represent the same function (only for different spatial locations), they represent the same concept [3]. For our purposes, we refer to the output of a convolutional filter after the activation function as a unit. We denote the output activations of the final convolutional layer of the network by the tensor $A \in \mathbb{R}^{H' \times W' \times K}$ where $K$ represents the number of channels in that layer. After computing the distribution of individual unit activations $a_k$, we determine the top quantile level $\mathcal{T}_k$ for each unit $k$ such that $P(a_k > \mathcal{T}_k) = 0.005$ [3]. We then derive the binary segmentation mask $M_k(\boldsymbol{x}) := A_k(\boldsymbol{x}) > \mathcal{T}_k$ and denote the set of enabled units for an input $\boldsymbol{x}$ as $E_x := \{k \mid \sum M_k(\boldsymbol{x}) > 0\}$.

**Positive Prediction Correlation.** Some units might capture concepts that are highly useful to determine whether a sample is fractured, establishing a stronger correlation with a true positive prediction than other units. To find these units, we compute:

$$c_k := \frac{\sum_{x \in P} \mathbf{1}_{E_x}(k)}{|P|} \tag{1}$$

where $P$ is the set of positive samples and $\mathbf{1}$ is the indicator function. With $c_{k_1} > c_{k_2} > ...$, $k_1$ is the unit most strongly correlated with a true positive prediction, followed by $k_2$.

### 2.3   Visualization of Highly Correlating Concepts at Inference

Due to the variability of observed defects in fractured vertebrae, different concepts are relevant during the inference of a sample. We compute the relevance of a unit $k$ during inference of input $\boldsymbol{x}$ as follows:

$$r_k := \sum M_k(\boldsymbol{x}) \odot A_k(\boldsymbol{x}) \tag{2}$$

For units $k_1$, $k_2$ with $r_{k_1} > r_{k_2}$, $k_1$ is more relevant for the inference of $\boldsymbol{x}$ than $k_2$. Now, when visualizing highly correlated concepts for a sample $\boldsymbol{x}$, we compute the inference relevance of each detector unit and display the activation maps $A_{k_1}(x)$, $A_{k_2}(x)$, ... with $r_{k_1} > r_{k_2} > ...$, showing the corresponding responses for the input sample $\boldsymbol{x}$.

## 3   Experimental Setup

**Data Preparation.** The network is trained on the VerSe dataset [29] as well as an in-house dataset acquired at Hospital A and Hospital B. The latter includes 465 patients with a median age of $\sim 69(\pm 12)$ years, containing a heterogeneous collection of field of views, scanner settings, and healthy and fractured vertebra, including metallic implants and foreign materials. This combined dataset contains CT scans of patients with healthy and fractured vertebrae of osteoporotic or malignant nature from a heterogeneous collection of CT scanners. To address the inherent class imbalance in the data, negative samples are undersampled and positive (fractured) samples are oversampled in training to achieve a perfect class balance each epoch. As osteoporotic and malignant fractures rarely occur in cervical vertebrae (C1–C7), they are excluded from the dataset. We extract $96 \times 96 \times 96$ sized 3D patches for each vertebrae with a 1mm resolution. These patches are centered on the vertebral body and oriented along the spine by aligning the vertical axis with a spline constructed with the vertebral centroids provided by the dataset similar to [15]. The intensity values of the resulting crops are cropped to a Hounsfield Unit range of $[-1000, 1000]$ and then scaled to $[0, 1]$. During training, intensity (Gaussian noise, smoothing, and contrast) and heavy spatial data augmentations (similarity transformation and elastic deformation) are applied. For these tasks, NiBabel 3.2.1 and MONAI 0.8.0 are used.

**Implementation Details.** The 3D U-Net is implemented in PyTorch Lightning 1.5.10 on top of PyTorch 1.10.2, and trained using the Adam [21] optimizer (learning rate 0.001) without weight decay. Training is concluded if the validation F1 score has not improved for 50 epochs. Dropout with probability 0.3 is applied.

## 4  Results and Discussion

In the following, we first evaluate the performance of our vertebral fracture detection neural network before dissecting it into its individual detector units. We then validate detector units highly correlated with a true positive prediction by showing that they represent clinically meaningful concepts. Lastly, we present a system to display the units most relevant to a single inference.

**Vertebral Fracture Detection.** We consider the threshold-based evaluation metrics F1-score and accuracy, evaluated at the vertebra level. To remove the dependence on a manually chosen threshold whose optimum might vary between trained networks, the area under curve (AUC) and average precision (AP) metrics are also evaluated. We report the mean and standard deviation of these metrics from five separate training trials for each model (Table 1).

**Table 1.** Performance of the trained neural networks on the test holdout of the smaller VerSe dataset as well as the combined dataset, comprised of VerSe and non-public data acquired from Hospital A and Hospital B. In total, the VerSe dataset contains 3,920 non-cervical vertebrae (254 of which are fractured), whereas the combined dataset comprises 10,675 T1-L5 vertebrae (1,246 fractured).

| Training | Testing | F1 (%) | Acc. (%) | AUC (%) | AP (%) |
|---|---|---|---|---|---|
| VerSe | VerSe | $71.2 \pm 10.8$ | $78.2 \pm 12.0$ | $84.5 \pm 9.1$ | $76.4 \pm 14.5$ |
| VerSe, in-house | VerSe | $86.1 \pm 2.6$ | $\mathbf{90.9 \pm 1.6}$ | $\mathbf{96.2 \pm 0.9}$ | $94.1 \pm 1.6$ |
| VerSe, in-house | VerSe, in-house | $\mathbf{88.0 \pm 0.7}$ | $88.0 \pm 0.4$ | $94.7 \pm 0.5$ | $\mathbf{95.0 \pm 0.4}$ |

For networks trained on the smaller VerSe dataset, we observe performance akin to "naive" two-dimensional vertebral fracture detection approaches on the same dataset [15], and a high dependence on a beneficial random seed. These networks, however, do not yield detector units that exhibit any discerning patterns. This is achieved by training a network with the larger dataset, combining VerSe and in-house data collected at Hospital A and Hospital B, that is reliably superior in performance. Its detector units exhibit a variety of patterns that are investigated in the subsequent sections.

**Table 2.** Visualization of the detector units most strongly correlated with a true positive prediction along with an interpretation of their activations by clinical experts. All displayed samples are fractured and represented by a slice with high activation after thresholding.

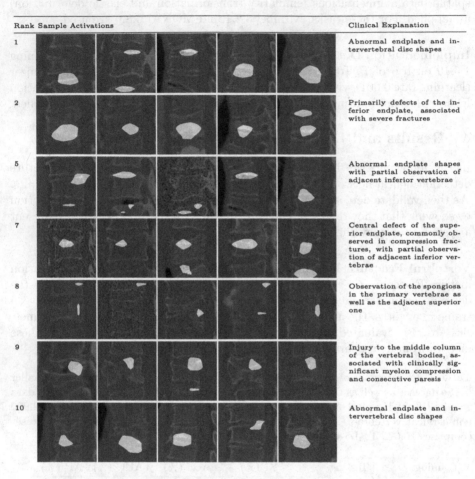

| Rank | Sample Activations | Clinical Explanation |
|---|---|---|
| 1 | | Abnormal endplate and intervertebral disc shapes |
| 2 | | Primarily defects of the inferior endplate, associated with severe fractures |
| 5 | | Abnormal endplate shapes with partial observation of adjacent inferior vertebrae |
| 7 | | Central defect of the superior endplate, commonly observed in compression fractures, with partial observation of adjacent inferior vertebrae |
| 8 | | Observation of the spongiosa in the primary vertebrae as well as the adjacent superior one |
| 9 | | Injury to the middle column of the vertebral bodies, associated with clinically significant myelon compression and consecutive paresis |
| 10 | | Abnormal endplate and intervertebral disc shapes |

## 4.1  Clinical Meaningfulness of Extracted Semantic Concepts

Given the network trained on the larger dataset, we extract its semantic concepts with Network Dissection [3], which we extended to the three-dimensional space. To reduce the 512 detector units of the 3D U-Net to a tractable number, we determine the top ten units highly correlated with a true positive prediction as detailed in Sect. 2.2. For these units, we exported a single-slice collage of 25 strongly activating fractured samples serving as an overview of the units' activations. For the five samples that activated the unit most strongly, all two-dimensional slices as well as three-dimensional NIfTI files are exported, allowing for a detailed inspection.

Based on these exports, we consulted two clinical experts with a combined experience of 22 years in spine imaging about the clinical meaningfulness of these detector units. Omitting three units where no immediate association was possible, we show the detector units identified by their correlation rank with their corresponding clinical explanation in Table 2. The provided samples show a diverse collection of detector unit activations, with each unit exhibiting consistent patterns across multiple samples. We also observe that these units' main focus is the primary vertebra, even if there is some activation in the surroundings. It is noteworthy that the patterns align with the bone anatomy and present themselves in clinically significant locations. As severe fractures are associated with changes in the superior and inferior vertebral endplates, we find the majority of activations in these regions. Although multiple detector units target these areas, they focus on different locations and exhibit varying sizes of regions of interest, with some integrating further information from the intervertebral discs as well as the adjacent vertebra. These insights are clinically meaningful to detect moderate and severe vertebral deformations (Genant grade 1 or higher [11]), and thus show that our network learned concepts that have a clinical correspondence.

For the omitted cases, we observed either no statistically significant activations, i.e. $M_k(x) = 0$, or sporadic activations that do not present any clear patterns, even though they are highly correlated with a true positive prediction. Overall, such detector units represent a minority and can therefore be disregarded in light of those that exhibit tangible patterns.

## 4.2   Single-Inference Concept Visualization

Having shown that the network learns clinically relevant concepts, we have validated its ability to make use of conducive features. We further seek to illuminate the black box decision-making process of the network by providing the user with a visual explanation for a single inference. To this end, we propose a system that visualizes the concepts considered most important by the network during inference.

Using the method described in Sect. 2.3 to identify the units representing the most relevant concepts, we retrieve their respective top activating images from our combined dataset. We then display two visualizations for each unit: (i) the activations of those units for the input sample, and (ii) the activations for their corresponding top images. This provides the user with a detector unit's particular response for the given input sample as well as a larger context to understand its general concept. For both visualizations, a single slice with high activation (after thresholding) is shown. An example of (i) is given with Table 3, which gives evidence of the network corroborating its prediction with a diverse set of concepts. These concepts illustrate the network accurately identifying relevant indications for the wedge-shaped deformity and incorporating information from an adjacent vertebra.

This system enables users to comprehend the network's decision making, increasing trust in the system and allowing them to identify failure cases more easily. Furthermore, this approach does not require any prior concept matching

**Table 3.** Visualization of the most relevant detector units during class prediction of the sample shown on the left, which the network correctly predicted as fractured. Each detector unit is represented by a single slice activation for that particular sample. We also show its ranking in units highly correlated with a true positive prediction. We observe that the network uses concepts associated with wedge-shaped deformity and incorporates information from an adjacent vertebra

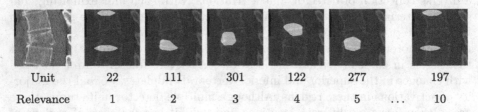

| Unit      | 22 | 111 | 301 | 122 | 277 | ... | 197 |
|-----------|----|-----|-----|-----|-----|-----|-----|
| Relevance | 1  | 2   | 3   | 4   | 5   | ... | 10  |

by experts, as the user is able to interpret the general concept of a detector unit and make informed judgements about its importance for a particular sample.

## 5    Conclusion

We show that a 3D U-Net learns a diverse set of concepts to tackle the task of detecting vertebral fractures. To gauge their meaningfulness, we first proposed a method to identify units highly correlated with a fracture detection. Then, we showed the overlap of these units with clinical concepts as validated by experts. Finally, we introduced a system to visually explain a single inference by showing the concepts most relevant for the classification of the sample, giving users insight into the network's decision making process. Further extensions of this system are conceivable, such as pre-filling a radiology report based on activations in a group of semantically similar detector units.

**Acknowledgements.** The authors acknowledge the financial support by the Federal Ministry of Education and Research of Germany (BMBF) under project DIVA (FKZ 13GW0469C). Ashkan Khakzar was partially supported by the Munich Center for Machine Learning (MCML) with funding from the BMBF under the project 01IS18036B. Kristina Mach was partially supported by the Linde & Munich Data Science Institute, Technical University of Munich PhD Fellowship.

## References

1. Bar, A., Wolf, L., Amitai, O.B., Toledano, E., Elnekave, E.: Compression fractures detection on CT. In: Medical Imaging 2017: Computer-Aided Diagnosis, vol. 10134, p. 1013440. International Society for Optics and Photonics (2017)
2. Bartalena, T., et al.: Prevalence of thoracolumbar vertebral fractures on multidetector CT. Eur. J. Radiol. **69**(3), 555–559 (2009)

3. Bau, D., Zhou, B., Khosla, A., Oliva, A., Torralba, A.: Network dissection: quantifying interpretability of deep visual representations. In: Proceedings of the IEEE Conference on Computer Vision and Pattern Recognition, pp. 6541–6549 (2017)
4. Bliuc, D.: Mortality risk associated with low-trauma osteoporotic fracture and subsequent fracture in men and women. JAMA 301(5), 513 (2009). https://doi.org/10.1001/jama.2009.50
5. Cauley, J.A.: Public health impact of osteoporosis. J. Gerontol. A Biol. Sci. Med. Sci. 68(10), 1243–1251 (2013)
6. Cauley, J., Thompson, D., Ensrud, K., Scott, J., Black, D.: Risk of mortality following clinical fractures. Osteoporos. Int. 11(7), 556–561 (2000)
7. Center, J.R., Nguyen, T.V., Schneider, D., Sambrook, P.N., Eisman, J.A.: Mortality after all major types of osteoporotic fracture in men and women: an observational study. The Lancet 353(9156), 878–882 (1999)
8. Chettrit, D., et al.: 3D convolutional sequence to sequence model for vertebral compression fractures identification in CT. In: Martel, A.L., et al. (eds.) MICCAI 2020. LNCS, vol. 12266, pp. 743–752. Springer, Cham (2020). https://doi.org/10.1007/978-3-030-59725-2_72
9. Çiçek, Ö., Abdulkadir, A., Lienkamp, S.S., Brox, T., Ronneberger, O.: 3D U-Net: learning dense volumetric segmentation from sparse annotation. In: Ourselin, S., Joskowicz, L., Sabuncu, M.R., Unal, G., Wells, W. (eds.) MICCAI 2016. LNCS, vol. 9901, pp. 424–432. Springer, Cham (2016). https://doi.org/10.1007/978-3-319-46723-8_49
10. Feng, S., Liu, B., Zhang, Y., Zhang, X., Li, Y.: Two-stream compare and contrast network for vertebral compression fracture diagnosis. IEEE Trans. Med. Imaging 40(9), 2496–2506 (2021)
11. Genant, H.K., Wu, C.Y., Van Kuijk, C., Nevitt, M.C.: Vertebral fracture assessment using a semiquantitative technique. J. Bone Miner. Res. 8(9), 1137–1148 (1993)
12. Haczynski, J., Jakimiuk, A.: Vertebral fractures: a hidden problem of osteoporosis. Med. Sci. Monit. Int. Med. J. Exp. Clin. Res. 7(5), 1108–1117 (2001)
13. Hallberg, I., Bachrach-Lindström, M., Hammerby, S., Toss, G., Ek, A.C.: Health-related quality of life after vertebral or hip fracture: a seven-year follow-up study. BMC Musculoskelet. Disord. 10(1), 135 (2009). https://doi.org/10.1186/1471-2474-10-135
14. Hernlund, E., et al.: Osteoporosis in the European Union: medical management, epidemiology and economic burden: a report prepared in collaboration with the International Osteoporosis Foundation (IOF) and the European Federation of Pharmaceutical Industry Associations (EFPIA). Arch. Osteoporos. 8(1-2), 136 (2013). https://doi.org/10.1007/s11657-013-0136-1
15. Husseini, M., Sekuboyina, A., Loeffler, M., Navarro, F., Menze, B.H., Kirschke, J.S.: Grading loss: a fracture grade-based metric loss for vertebral fracture detection. In: Martel, A.L., et al. (eds.) MICCAI 2020. LNCS, vol. 12266, pp. 733–742. Springer, Cham (2020). https://doi.org/10.1007/978-3-030-59725-2_71
16. Jalava, T., et al.: Association between vertebral fracture and increased mortality in osteoporotic patients. J. Bone Miner. Res. 18(7), 1254–1260 (2003). https://doi.org/10.1359/jbmr.2003.18.7.1254
17. Khakzar, A., Baselizadeh, S., Navab, N.: Rethinking positive aggregation and propagation of gradients in gradient-based saliency methods. arXiv preprint arXiv:2012.00362 (2020)
18. Khakzar, A., Khorsandi, P., Nobahari, R., Navab, N.: Do explanations explain? Model knows best. arXiv preprint arXiv:2203.02269 (2022)

19. Khakzar, A., et al.: Towards semantic interpretation of thoracic disease and COVID-19 diagnosis models. In: de Bruijne, M., et al. (eds.) MICCAI 2021. LNCS, vol. 12903, pp. 499–508. Springer, Cham (2021). https://doi.org/10.1007/978-3-030-87199-4_47

20. Khakzar, A., et al.: Explaining COVID-19 and thoracic pathology model predictions by identifying informative input features. In: de Bruijne, M., et al. (eds.) MICCAI 2021. LNCS, vol. 12903, pp. 391–401. Springer, Cham (2021). https://doi.org/10.1007/978-3-030-87199-4_37

21. Kingma, D.P., Ba, J.: Adam: a method for stochastic optimization. arXiv preprint arXiv:1412.6980 (2014)

22. Li, Y., et al.: Differential diagnosis of benign and malignant vertebral fracture on CT using deep learning. Eur. Radiol. **31**(12), 9612–9619 (2021)

23. Lundberg, S.M., Lee, S.I.: A unified approach to interpreting model predictions. In: Advances in Neural Information Processing Systems (2017)

24. Melton, L.J., III., Atkinson, E.J., Cooper, C., O'Fallon, W.M., Riggs, B.L.: Vertebral fractures predict subsequent fractures. Osteoporos. Int. **10**(3), 214–221 (1999). https://doi.org/10.1007/s001980050218

25. Montavon, G., Lapuschkin, S., Binder, A., Samek, W., Müller, K.R.: Explaining nonlinear classification decisions with deep Taylor decomposition. Pattern Recogn. (2017). https://doi.org/10.1016/j.patcog.2016.11.008

26. Nicolaes, J., et al.: Detection of vertebral fractures in CT using 3D convolutional neural networks. In: Cai, Y., Wang, L., Audette, M., Zheng, G., Li, S. (eds.) CSI 2019. LNCS, vol. 11963, pp. 3–14. Springer, Cham (2020). https://doi.org/10.1007/978-3-030-39752-4_1

27. Pisov, M., et al.: Keypoints localization for joint vertebra detection and fracture severity quantification. In: Martel, A.L., et al. (eds.) MICCAI 2020. LNCS, vol. 12266, pp. 723–732. Springer, Cham (2020). https://doi.org/10.1007/978-3-030-59725-2_70

28. Schulz, K., Sixt, L., Tombari, F., Landgraf, T.: Restricting the flow: information bottlenecks for attribution. In: International Conference on Learning Representations (2020). https://openreview.net/forum?id=S1xWh1rYwB

29. Sekuboyina, A., et al.: Verse: a vertebrae labelling and segmentation benchmark for multi-detector CT images. Med. Image Anal. **73**, 102166 (2021)

30. Tomita, N., Cheung, Y.Y., Hassanpour, S.: Deep neural networks for automatic detection of osteoporotic vertebral fractures on CT scans. Comput. Biol. Med. **98**, 8–15 (2018)

31. Williams, A.L., Al-Busaidi, A., Sparrow, P.J., Adams, J.E., Whitehouse, R.W.: Under-reporting of osteoporotic vertebral fractures on computed tomography. Eur. J. Radiol. **69**(1), 179–183 (2009)

32. Wu, J., et al.: Deepminer: discovering interpretable representations for mammogram classification and explanation. arXiv preprint arXiv:1805.12323 (2018)

33. Yilmaz, E.B., et al.: Automated deep learning-based detection of osteoporotic fractures in CT images. In: Lian, C., Cao, X., Rekik, I., Xu, X., Yan, P. (eds.) MLMI 2021. LNCS, vol. 12966, pp. 376–385. Springer, Cham (2021). https://doi.org/10.1007/978-3-030-87589-3_39

34. Yilmaz, E.B., Mader, A.O., Fricke, T., Peña, J., Glüer, C.-C., Meyer, C.: Assessing attribution maps for explaining CNN-based vertebral fracture classifiers. In: Cardoso, J., et al. (eds.) IMIMIC/MIL3ID/LABELS -2020. LNCS, vol. 12446, pp. 3–12. Springer, Cham (2020). https://doi.org/10.1007/978-3-030-61166-8_1

35. Zhang, Y., Khakzar, A., Li, Y., Farshad, A., Kim, S.T., Navab, N.: Fine-grained neural network explanation by identifying input features with predictive information. In: Advances in Neural Information Processing Systems, vol. 34 (2021)
36. Zhou, B., Khosla, A., Lapedriza, A., Oliva, A., Torralba, A.: Learning deep features for discriminative localization. In: Proceedings of the IEEE Conference on Computer Vision and Pattern Recognition, pp. 2921–2929 (2016)

# Multi-modal Volumetric Concept Activation to Explain Detection and Classification of Metastatic Prostate Cancer on PSMA-PET/CT

R. C. J. Kraaijveld[1], M. E. P. Philippens[2], W. S. C. Eppinga[2],
I. M. Jürgenliemk-Schulz[2], K. G. A. Gilhuijs[1], P. S. Kroon[2],
and B. H. M. van der Velden[1(✉)]

[1] Image Sciences Institute, University Medical Center Utrecht,
Utrecht, The Netherlands
bvelden2@umcutrecht.nl
[2] Department of Radiotherapy, University Medical Center Utrecht,
Utrecht, The Netherlands

**Abstract.** Explainable artificial intelligence (XAI) is increasingly used to analyze the behavior of neural networks. Concept activation uses human-interpretable concepts to explain neural network behavior. This study aimed at assessing the feasibility of regression concept activation to explain detection and classification of multi-modal volumetric data.

Proof-of-concept was demonstrated in metastatic prostate cancer patients imaged with positron emission tomography/computed tomography (PET/CT). Multi-modal volumetric concept activation was used to provide global and local explanations.

Sensitivity was 80% at 1.78 false positive per patient. Global explanations showed that detection focused on CT for anatomical location and on PET for its confidence in the detection. Local explanations showed promise to aid in distinguishing true positives from false positives. Hence, this study demonstrated feasibility to explain detection and classification of multi-modal volumetric data using regression concept activation.

**Keywords:** Explainable artificial intelligence · Interpretable deep learning · Medical image analysis · Prostate cancer · PET/CT

## 1 Introduction

Deep learning has revolutionized medical image analysis. The neural networks used in deep learning typically consist of many layers connected via many non-linear intertwined connections. Even if one was to inspect all these layers and connections, it is impossible to fully understand how the neural network reached its decision [17]. Hence, deep learning is often regarded as a 'black box' [17]. In high-stakes decision-making such as medical applications, this can have far-reaching consequences [18].

© The Author(s), under exclusive license to Springer Nature Switzerland AG 2022
M. Reyes et al. (Eds.): iMIMIC 2022, LNCS 13611, pp. 82–92, 2022.
https://doi.org/10.1007/978-3-031-17976-1_8

Medical experts have voiced their concern about this black box nature, and called for approaches to better understand the black box [11]. Such approaches are commonly referred to as interpretable deep learning or explainable artificial intelligence (XAI) [1]. Visual explanation is the most frequently used XAI [21]. There is increasing evidence that the saliency maps that provide this visual explanation are to be used with caution [2,3,6]. For example, they can be incorrect and not correspond to what the end-user expected from the explanation (i.e., low validity) or lack robustness [21]. Hence, such methods may not be as interpretable as desired.

In response to "uninterpretable" XAI, Kim et al. proposed to use human-interpretable concepts for explaining models (e.g. a neural network) [12]. Examples of such concepts are a spiculated tumor margin – a sign of malignant breast cancer [8] – or the short axis of a metastatic lymph node in a prostate cancer patient, which has been related to patient prognosis [16]. Using concepts, Kim et al. were able to test how much a concept influenced the decision of the model (i.e., concept activation) [12].

Concept activation has been used in medical image analysis to explain classification techniques using binary concepts [12] – such as the presence of micro-aneurysms in diabetic retinopathy – and continuous concepts (i.e., regression concept activation) [9] – such as the area of nuclei in breast histopathology. To the best of our knowledge, the promise of concept activation has not yet been shown in detection, 3-dimensional volumetric data, or multi-modal data.

The aim of this study was to assess the feasibility of regression concept activation to explain detection and classification of multi-modal volumetric data. We demonstrated proof-of-concept in patients who had metastatic prostate cancer.

## 2  Data

A total of 88 consecutively included male patients with oligometastatic (i.e., five or less metastatic lymph nodes) prostate cancer from the University Medical Center Utrecht were analysed. All patients gave written informed consent and the study was approved by the local medical ethics committee [22]. Median age was 71 years with an interquartile interval of 67–74 years.

Patients were imaged using $^{68}$Ga prostate-specific membrane antigen positron emission tomography and computed tomography (PSMA-PET/CT) (Fig. 1). The in-plane voxel size of the PET scans ranged from $1.5\,mm^2$ to $4.1\,mm^2$, slice thickness ranged from $1.5\,mm$ to $5.0\,mm$. The in-plane voxel size of the CT scans ranged from $0.84\,mm^2$ to $1.4\,mm^2$, slice thickness was $2.0\,mm$.

Metastatic lymph nodes were delineated by a radiation oncologist in consensus with a nuclear medicine physician. Furthermore, lymph nodes were confirmed on magnetic resonance imaging.

## 3  Method

In short, we first detected the metastases and subsequently filtered out false positive detections at high sensitivity using classification. XAI was used on both the detection and the classification to provide global and local explanation (Fig. 2).

PSMA-PET (MIP)

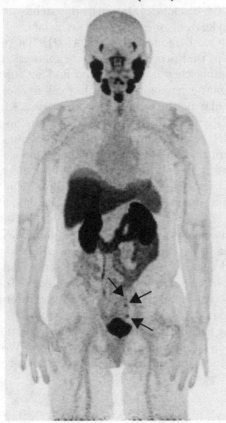

PSMA-PET (ROI slice)

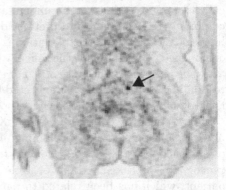

CT (ROI slice)

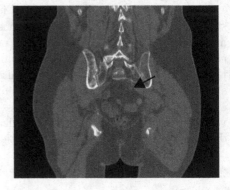

**Fig. 1.** Example of a prostate cancer patient with three metastatic lymph nodes. Left: maximum intensity projection (MIP) of prostate-specific membrane antigen positron emission tomography (PSMA-PET) showing three metastatic lymph nodes. Right: region of interest (ROI) showing one of the metastatic lymph nodes on PSMA-PET and on computed tomography (CT).

## 3.1   Preprocessing

PET scans were registered to the CT scans. Data was split into 70 patients for training/validation and 18 patients for testing. This resulted in 109 metastatic lymph nodes for training and 30 for testing.

## 3.2   Detection

nnDetection [4] was used to detect the metastatic lymph nodes. Input to nnDetection were PET/CT images, output were 3D bounding boxes with corresponding intersection-over-union and confidence scores. Hyperparameters were optimized by nnDetection.

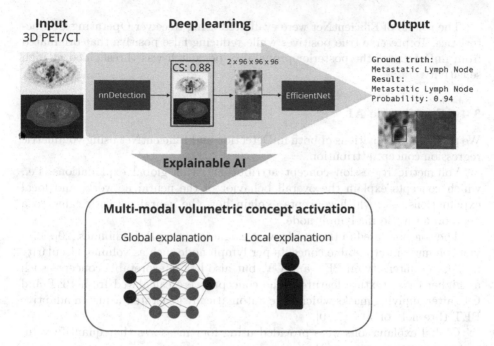

**Fig. 2.** Schematic overview of the method. First, nnDetection detects metastatic lymph nodes on multi-modal volumetric positron emission tomography and computed tomography (PET/CT) images. These detections are then refined using EfficientNet. An XAI – multi-modal volumetric concept activation – is used to provide global and local explanations. CS = confidence score.

The results of nnDetection were evaluated using Free-response Receiver Operating Characteristics. To ensure high metastatic lymph node detection rate, the intersection-over-union and confidence scores were thresholded at high sensitivity.

## 3.3   Classification

EfficientNet [19] was used to subsequently filter out false positive detections by classifying bounding boxes originating from nnDetection. PET/CT volumes of $96 \times 96 \times 96$ (i.e., patches) were extracted. These patches were input to Efficient-Net, output were binary classes representing whether there was a metastatic lymph node present or not. EfficientNet was trained using Adam optimizer and cross entropy loss. The initial learning rate was set as 0.001 and decreased step-wise by 0.10 every 5 epochs. EfficientNet was trained for 25 epochs with early-stopping. Augmentation included horizontal and vertical flipping, translation, scaling and rotation. Weighted random sampling was used to minimize the effect of class imbalance.

The results of EfficientNet were evaluated using Receiver Operating Characteristics. To preserve true positives while reducing false positive that originated from nnDetection, the posterior probability per patch was thresholded at high sensitivity.

### 3.4   Explainable AI

We provided explanations of both nnDetection and EfficientNet using volumetric regression concept attribution.

Volumetric regression concept attribution yields global explanations, i.e., which concepts explain the overall behavior of the neural network, and local explanations, i.e., which concepts explain how the neural network came to a decision for a specific lymph node.

The concepts used in this study were extracted using PyRadiomics [20]. This yields human-interpretable concepts per lymph node such as volume, circularity in 3D, and intensity on PET and CT, but also less interpretable concepts such as higher order texture features. The concepts were calculated from PET and CT, after applying masks which were automatically generated using an adaptive PET threshold of 40% [7,10].

Global explanations were provided using four measures that quantify volumetric regression concept attribution:

1. Pearson's correlation coefficient $\rho$ was calculated between each feature and either the confidence scores in case of nnDetection or the posterior probability in case of EfficientNet.
2. The regression coefficient and regression concept vector were assessed per feature by fitting a linear model between layer activations and feature values. For each layer in the neural network, a regression coefficient can be quantified per concept, revealing the learning behavior of the neural network.
3. Sensitivity scores were calculated which indicate the influence of the concept on the outcome of the neural network result.
4. The bidirectional relevance was calculated for each concept by taking the product of the regression coefficient and the inverse of the coefficient of variation of the sensitivity scores.

Local explanations were provided by comparing the sensitivity score of a concept per input image to the mean sensitivity of that concept. The difference between these sensitivity scores can be used as a similarity measure of that input image to an output class (e.g., metastatic lymph node).

**Computation:** Deep learning was done in PyTorch 1.8 on an NVIDIA GeForce 2080Ti. Code will be available at https://github.com/basvandervelden/mmvca.

# 4 Results

## 4.1 Detection

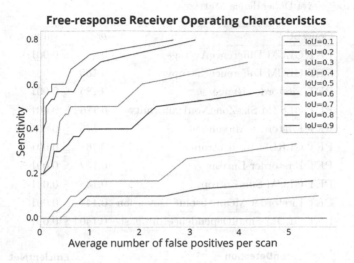

**Fig. 3.** At an intersection-over-union (IoU) of 0.1, 0.80 sensitivity was obtained at 2.66 false positives per patient (top line).

At an intersection-over-union of 0.1, a sensitivity of 0.80 was obtained at an average of 2.66 false positive per patient (Fig. 3). In total, 24 out of 30 lymph nodes were detected at the cost of 48 false positives.

## 4.2 Classification

EfficientNet showed an additional reduction of 16 of the 48 false positives that originated from nnDetection (33% reduction), while maintaining all true positives. Hence, the final amount of false positives per patient was 1.78.

## 4.3 Explainable AI

**Global Explanations:** Table 1 shows the top ten concepts with the highest Pearson's correlation coefficient $\rho$ between the concepts and confidence scores of the bounding boxes from nnDetection. All these top ten concepts originate from the PET scan. Figure 4 shows the top ten bidirectional relevance scores for nnDetection. All these top ten concepts originate from the CT scan.

Table 2 shows the top ten concepts with the highest Pearson's correlation coefficient $\rho$ between the concept and the posterior probability of a metastatic lymph node in the patch. Figure 4 shows which concepts influence the classification results the most. These top ten concepts for both XAI measures originate from the PET scan.

**Table 1.** All of the top ten correlations between concepts and the confidence scores of the bounding boxes originate from the positron emission tomography (PET) scan. GLCM = Gray Level Cooccurence Matrix, First order = First order statistics, GLSZM = Gray Level Size Zone Matrix, GLRLM = Gray Level Run Length Matrix, GLDM = Gray Level Dependence Matrix.

| Concept | $\rho$ | P-value |
|---|---|---|
| PET GLCM DifferenceAverage | 0.186 | $\leqq 0.001$ |
| PET GLCM DifferenceEntropy | 0.185 | $\leqq 0.001$ |
| PET Firstorder Range | 0.185 | $\leqq 0.001$ |
| PET GLSZM SizeZoneNonUniformity | 0.176 | $\leqq 0.001$ |
| PET Firstorder Maximum | 0.175 | $\leqq 0.001$ |
| PET GLRLM RunEntropy | 0.168 | $\leqq 0.001$ |
| PET Firstorder Entropy | 0.152 | $\leqq 0.001$ |
| PET GLCM SumEntropy | 0.148 | $\leqq 0.001$ |
| PET Firstorder MeanAbsoluteDeviation | 0.147 | $\leqq 0.001$ |
| PET GLDM SmallDependenceEmphasis | 0.140 | $\leqq 0.001$ |

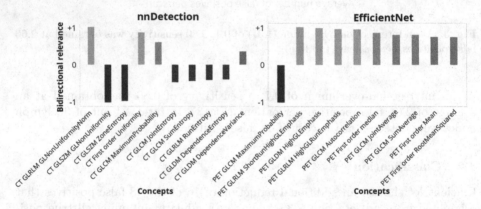

**Fig. 4.** The top ten concepts with the highest bidirectional relevance originate from the computed tomography (CT) scan for nnDetection (left) and from the positron emission tomography (PET) scan for EfficientNet (right). GL = Gray level, Norm = normalized, GLRLM = Gray Level Run Length Matrix, GLSZM = Gray Level Size Zone Matrix, First order = First order statistics, GLCM = Gray Level Cooccurence Matrix, GLDM = Gray Level Dependence Matrix. (Color figure online)

**Local Explanations:** Figure 5 shows how the local explanations can be used by a physician. Each case was ranked according to its similarity with a metastatic lymph node and its top ten concepts.

To further investigate the six undetected lymph nodes from nnDetection, we also evaluated these in a post hoc analysis with EfficientNet. Four of the six (66%) false negatives were correctly classified as a lymph node. Local explanations showed that the two incorrectly classified lymph nodes had low similarity with the class metastatic lymph node, according to the top ten concepts.

**Table 2.** All of the top ten correlations between concepts and the posterior probability of a metastatic lymph node in the patch originate from the positron emission tomography (PET) scan. First order = First order statistics, GLCM = Gray Level Cooccurence Matrix.

| Concept | $\rho$ | p-value |
|---|---|---|
| PET First order Range | 0.449 | $\leq 0.001$ |
| PET GLCM SumAverage | 0.444 | $\leq 0.001$ |
| PET GLCM JointAverage | 0.444 | $\leq 0.001$ |
| PET First order Median | 0.442 | $\leq 0.001$ |
| PET First order Maximum | 0.436 | $\leq 0.001$ |
| PET First order Mean | 0.430 | $\leq 0.001$ |
| PET First order RootMeanSquared | 0.429 | $\leq 0.001$ |
| PET GLCM MCC | 0.428 | $\leq 0.001$ |
| PET First order 10Percentile | 0.425 | $\leq 0.001$ |
| PET First order 90Percentile | 0.423 | $\leq 0.001$ |

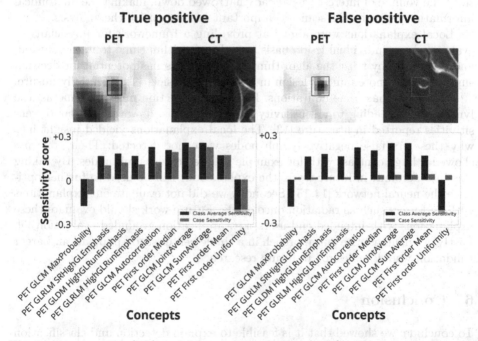

**Fig. 5.** True positive (left) and false positive finding (right) with their corresponding local explanation underneath. It can be seen that the sensitivity scores of the left PET/CT patch reflects the class sensitivity scores. In the right PET/CT patch the sensitivity scores differ substantially from the class sensitivity scores. Hence, this local explanation can give an extra confirmation to the physician to rule this a false positive. GLCM = Gray Level Cooccurence Matrix, GLRLM = Gray Level Run Length Matrix, GLDM = Gray Level Dependence Matrix, First order = First order statistics. (Color figure online)

# 5    Discussion

This study showed feasibility of regression concept activation to explain detection and classification of multi-modal volumetric data. In 88 oligometastatic prostate cancer patients, our method was able to provide realistic global and local explanations.

The global explanations for nnDetection yielded plausible results. Confidence scores of nnDetection's bounding boxes were all positively correlated with concepts from the PET scan, whereas the concepts that influenced the position of the bounding boxes came from the CT scan. In other words, the CT scan provides detailed anatomical information explaining in which region of the patient lymph nodes could be present, whereas the PET scan influences how confident the network is that the detection is actually a metastatic lymph node. Since PSMA-PET is designed for this specific goal, these explanations are plausible.

The global explanations for EfficientNet also yielded plausible results. The posterior probability whether a metastatic lymph node was present in a patch was mostly correlated with concepts from the PET scan. This again makes sense, since the volume of interest was already narrowed down, making the anatomical information from the CT scan less important in this part of the analysis.

Local explanations were aimed at providing a framework for physicians to evaluate on an individual lesion basis how the algorithm came to its conclusion, and whether they trust the algorithms decision. This has potential for decision support in the more difficult lesion in which the physician is potentially unsure.

This study has some limitations. Firstly, nnDetection misses six metastatic lymph nodes, leading to a sensitivity of 80%. This is, however, similar to sensitivities reported in literature [13]. The local explanations yielded insight into why these six false negative lymph nodes were not detected: Their concepts showed a large contrast with for example the detected lymph nodes. By taking this into account, in future work, the explanations can be used to further optimize the neural network [14,15]. Secondly, we did not evaluate our explanations with end-users such as radiation oncologists. Future work should evaluate these explanations with intended end-users, i.e., application-grounded evaluation [5]. Lastly, we demonstrate our approach in a single center study population. Larger validation would be desired in future research.

# 6    Conclusion

To conclude, we showed that it is feasible to explain detection and classification of multi-modal volumetric data using regression concept activation.

# References

1. Adadi, A., Berrada, M.: Peeking inside the black-box: a survey on explainable artificial intelligence (XAI). IEEE Access **6**, 52138–52160 (2018)

2. Adebayo, J., Gilmer, J., Muelly, M., Goodfellow, I., Hardt, M., Kim, B.: Sanity checks for saliency maps. In: Advances in Neural Information Processing Systems, vol. 31 (2018)
3. Arun, N., et al.: Assessing the trustworthiness of saliency maps for localizing abnormalities in medical imaging. Radiol. Artif. Intell. **3**(6) (2021)
4. Baumgartner, M., Jäger, P.F., Isensee, F., Maier-Hein, K.H.: nnDetection: a self-configuring method for medical object detection. In: de Bruijne, M., et al. (eds.) MICCAI 2021. LNCS, vol. 12905, pp. 530–539. Springer, Cham (2021). https://doi.org/10.1007/978-3-030-87240-3_51
5. Doshi-Velez, F., Kim, B.: Towards a rigorous science of interpretable machine learning. arXiv preprint arXiv:1702.08608 (2017)
6. Eitel, F., Ritter, K.: Testing the robustness of attribution methods for convolutional neural networks in MRI-based Alzheimer's disease classification. In: Suzuki, K., et al. (eds.) ML-CDS/IMIMIC -2019. LNCS, vol. 11797, pp. 3–11. Springer, Cham (2019). https://doi.org/10.1007/978-3-030-33850-3_1
7. Erdi, Y.E., et al.: Segmentation of lung lesion volume by adaptive positron emission tomography image thresholding. Cancer Interdisc. Int. J. Am. Cancer Soc. **80**(S12), 2505–2509 (1997)
8. Gilhuijs, K.G., Giger, M.L., Bick, U.: Computerized analysis of breast lesions in three dimensions using dynamic magnetic-resonance imaging. Med. Phys. **25**(9), 1647–1654 (1998)
9. Graziani, M., Andrearczyk, V., Marchand-Maillet, S., Müller, H.: Concept attribution: explaining CNN decisions to physicians. Comput. Biol. Med. **123**, 103865 (2020)
10. Im, H.J., Bradshaw, T., Solaiyappan, M., Cho, S.Y.: Current methods to define metabolic tumor volume in positron emission tomography: which one is better? Nucl. Med. Mol. Imaging **52**(1), 5–15 (2018)
11. Jia, X., Ren, L., Cai, J.: Clinical implementation of AI technologies will require interpretable AI models. Med. Phys. **47**(1), 1–4 (2020)
12. Kim, B., Wattenberg, M., Gilmer, J., Cai, C., Wexler, J., Viegas, F., et al.: Interpretability beyond feature attribution: quantitative testing with concept activation vectors (TCAV). In: International Conference on Machine Learning, pp. 2668–2677. PMLR (2018)
13. Kim, S.J., Lee, S.W., Ha, H.K.: Diagnostic performance of radiolabeled prostate-specific membrane antigen positron emission tomography/computed tomography for primary lymph node staging in newly diagnosed intermediate to high-risk prostate cancer patients: a systematic review and meta-analysis. Urol. Int. **102**(1), 27–36 (2019)
14. Lund, C.B., van der Velden, B.H.M.: Leveraging clinical characteristics for improved deep learning-based kidney tumor segmentation on CT. In: Heller, N., Isensee, F., Trofimova, D., Tejpaul, R., Papanikolopoulos, N., Weight, C. (eds.) KiTS 2021. LNCS, vol. 13168, pp. 129–136. Springer, Cham (2022). https://doi.org/10.1007/978-3-030-98385-7_17
15. Mahapatra, D., Ge, Z., Reyes, M.: Self-supervised generalized zero shot learning for medical image classification using novel interpretable saliency maps. IEEE Trans. Med. Imaging (2022)
16. Meijer, H.J., Debats, O.A., van Lin, E.N., Witjes, J.A., Kaanders, J.H., Barentsz, J.O.: A retrospective analysis of the prognosis of prostate cancer patients with lymph node involvement on MR lymphography: who might be cured. Radiat. Oncol. **8**(1), 1–7 (2013)

17. Murdoch, W.J., Singh, C., Kumbier, K., Abbasi-Asl, R., Yu, B.: Definitions, methods, and applications in interpretable machine learning. Proc. Natl. Acad. Sci. **116**(44), 22071–22080 (2019)
18. Rudin, C.: Stop explaining black box machine learning models for high stakes decisions and use interpretable models instead. Nat. Mach. Intell. **1**(5), 206–215 (2019)
19. Tan, M., Le, Q.: Efficientnet: rethinking model scaling for convolutional neural networks. In: International Conference on Machine Learning, pp. 6105–6114. PMLR (2019)
20. Van Griethuysen, J.J., et al.: Computational radiomics system to decode the radiographic phenotype. Can. Res. **77**(21), e104–e107 (2017)
21. van der Velden, B.H., Kuijf, H.J., Gilhuijs, K.G., Viergever, M.A.: Explainable artificial intelligence (XAI) in deep learning-based medical image analysis. Med. Image Anal. 102470 (2022)
22. Werensteijn-Honingh, A.M., et al.: Progression-free survival in patients with 68Ga-PSMA-PET-directed SBRT for lymph node oligometastases. Acta Oncol. **60**(10), 1342–1351 (2021)

# KAM - A Kernel Attention Module for Emotion Classification with EEG Data

Dongyang Kuang[1](✉)[iD] and Craig Michoski[2][iD]

[1] School of Mathematics (Zhuhai), Sun Yat-sen University,
Zhuhai, Guangdong, China
dykuang@outlook.com
[2] Oden Institute for Computational Engineering and Sciences,
University of Texas at Austin, Austin, USA
michoski@oden.utexas.edu
https://dykuang.github.io/

**Abstract.** In this work, a kernel attention module is presented for the task of EEG-based emotion classification with neural networks. The proposed module utilizes a self-attention mechanism by performing a kernel trick, demanding significantly fewer trainable parameters and computations than standard attention modules. The design also provides a scalar for quantitatively examining the amount of attention assigned during deep feature refinement, hence help better interpret a trained model. Using EEGNet as the backbone model, extensive experiments are conducted on the SEED dataset to assess the module's performance on within-subject classification tasks compared to other SOTA attention modules. Requiring only one extra parameter, the inserted module is shown to boost the base model's mean prediction accuracy up to more than 1% across 15 subjects. A key component of the method is the interpretability of solutions, which is addressed using several different techniques, and is included throughout as part of the dependency analysis.

**Keywords:** Kernel attention · EEG · Emotion classification

## 1 Introduction

Correctly identifying human emotion using classification strategies has long been a topic of interest in brain computer interfaces (BCI) and their applications. According to the review [17] on classification algorithms utilized in EEG studies, there are five major categories of classifiers currently under investigation, which are: i) conventional classifiers [13–15,20], ii) matrix and tensor based classifiers [5], iii) transfer learning based methods [2,7], iv) deep learning algorithms and advanced statistical approaches [4,19], and v) multi-label classifiers [3,18,21]. While many classification approaches have been explored in the context of EEG signal processing, the classification pipeline itself has still frequently involved

M. Reyes et al. (Eds.): iMIMIC 2022, LNCS 13611, pp. 93–103, 2022.
https://doi.org/10.1007/978-3-031-17976-1_9

extensive manual preprocessing and feature engineering, often requiring intervention from domain experts as well as the experimental operators involved in the acquisition of the EEG signals themselves.

The success of neural network tools has been shown to alleviate some of limitations in the classical pipeline by providing, for example, faster predictions and reducing the need to manually preprocess data. One problem that remains however, is that the scale of learnable parameters in a classification network can be too large relative to the input data size. That is, in contrast to areas such as image classification, the availability of clean and open-sourced EEG data sets is comparably quite small in size. As a consequence, basic research calls specifically for data efficient and parameter efficient, as well as human interpretable models in order to provide penetrating insight into human emotion classification given relatively sparse data sampling.

## 2    Related Work

Below we briefly review some salient results in the literature relevant to the present work.

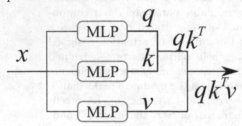

**Fig. 1.** The basic self-attention mechanism.

**Self-attention:** Self-attention was originally introduced in the field of natural language processing (NLP) [1], where the self-attention mechanism operates as a key component in transformer modules. These standard self-attention designs rely on MLPs, instead of the more conventionally used convolutional layers, to generate an attention matrix. Adapting the concept of self-attention to computer vision has been remarkably successful, and has demonstrated impressive performance to date on a variety of different tasks, e.g. [6,16]. As a consequence, through self-attention approaches, the field of image classification—long dominated by an assortment of solutions utilizing convolution neural networks—has discovered a promising new tool for (potentially) broad application.

**Alternative Attention Approaches:** In addition to self-attention, there exists many other types of attention modules in the field of computer vision. Two prominent examples of these are the Squeeze-and-Excitation (SE) network [10] and the Convolutional Block Attention Module (CBAM) [22], where the SE network won the ImageNet2017 championship, while CBAM sequentially infers attention maps along both channel and spatial dimensions for adaptive feature refinement. These methods, and others, are indicative of the increased contemporary importance attention-style approaches are having within computer vision.

**EEGNet:** EEGNet was proposed in [12] for a compact network design aimed at finding better generalizations across different BCI paradigms. Using depthwise/separable convolution layers, the EEGNet network contains considerably fewer trainable parameters than models constructed with regular convolutional layers, while still showing commensurate performance. EEGNet is also equipped with a convolutional layer of kernel size equal to the total number of EEG channels, enabling the ability to investigate spatial patterns learned during training in order to elucidate underlying electrophysical principles. Because of both the effectiveness and parsimonious nature of the EEGNet design, we adopt it as the backbone network for comparing and interpreting different attention modules.

**SEED:** The SEED dataset [23] includes multiple physiological signals that evaluate self-reported emotional responses, classified into *Positive, Neutral,* and *Negative* reactions taken from 15 participants, and is one of the standard datasets for benchmarking EEG signal classification strategies. The data was collected with 62 EEG channels using the 10–20 international standard.

**Our Work:** In this paper we consider a self-attention mechanism for boosting the backbone model's performance in a parameter efficient way and for providing better interpretation. However, at the outset, it is worth noting that several potential difficulties arise when conceptualizing the incorporation of a self-attention mechanism into the EEGNet framework. First, as discussed in [6], transformers tend to be quite data hungry models, failing to outperform regular convolution-based networks when the dataset's size is not large enough. In the area of emotion classification using EEG signals, this data-thirst requirement can become prohibitive. One of the reasons for this "data hungry" aspect of transformers is due to the MLP layers generally involved in the attention mechanism. These dense layers, not surprisingly, tend to contain many more parameters than convolutional layers that comprise popular emotion classification frameworks. Consequently, a difficulty arises when self-attention is applied directly to frameworks such as EEGNet, since the resulting hybrid frameworks tend to substantially undermine the primary advantage of the underlying base models; for example, EEGNet would no longer be a lightweight and compact model, but instead become a data hungry model focusing entirely on accuracy over pragmatic utility.

Thus, the primary goal of the present work is to find a way to incorporate the self-attention mechanism into EEGNet in a way that can still preserve EEGNet's pragmatic utility by maximizing the parameter efficiency in the design of the attention module. The solution presented in this paper is called a Kernel Attention network Module (KAM), and can be described as:

1. Utilizing a kernel function to produce the proper attention matrix instead of relying on MLP layers; thus reducing the number of both parameters and computations required. Moreover, benefiting from the one parameter design, specific techniques are then able to be employed for more effective interpretability techniques as well.
2. With the proposed module inserted along with only one additional parameter, the predictive performance of the baseline model—in our case, EEGNet—can be boosted up to more than 3% for some subjects on within-subject classifications and more than 1% overall on mean performance.

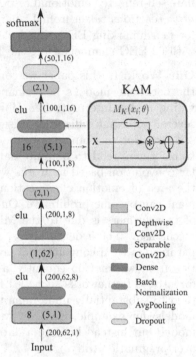

# 3   Kernel Attention Module

Figure 1 gives a simple illustration of how the basic self-attention mechanism works when applied on some feature $x$. First, the feature $x$ is mapped to three different features of the same size via $q = \phi_q(x)$, $k = \phi_k(x)$, and $v = \phi_v(x)$. The mappings $\phi_j$ for $j \in \{q, k, v\}$ are achieved using MLP blocks. Next, an attention matrix is formed by computing $qk^T$ which will then be used as a prefactor on $v$ to produce the attention output. Computationally, this procedure can be summarized as: $x \leftarrow [\phi_q(x)\phi_k(x)^T]\phi_v(x)$.

In the case where $x$ has large feature dimension $n$, the attention is applied on segments of $x$ in parallel with the resulting feature pieces subsequently concatenated afterwards—a mechanism referred to as *multi-head self-attention*. For more details on the underlying algorithms we refer the reader to [6]. Using the above basic self-attention format, KAM is constructed by replacing the inner product form $\phi_q(x)\phi_k(x)^T$ with a kernel matrix $M_K(x; \theta)$ subject to some parameter $\theta$. For example, a Gaussian type kernel function can be used to generate $M_K^{ij} = \exp(-\alpha d(x_i, x_j)^2)$, where $d(\cdot, \cdot)$ denotes some distance metric, $x_i$ is the $i$th row or column of feature block $x$ depending on whether $M_K$ is multiplied to $x$ by left or right, $\theta = \alpha$ is the learnable parameter during training. In the KAM design, $\phi_v$ can be simply dropped to reduce the number of total parameters. Finally, a skip connection is included in the KAM design that offers several potential benefits. On one hand, an additional skip can help better backpropagation of gradients to the blocks in front of the KAM. On the other, it provides an easy interface to quantitatively measure how much attention is actually being

**Fig. 2.** EEGNet with KAM inserted. Some important hyperparameters, kernel shapes and tensor sizes are also shown.

applied, requiring only an examination on the values of $\theta$. For example, when $\alpha \to +\infty$ then $M_K(x; \theta) \to I$, meaning no cross attention among features is applied. However, if $\alpha \to 0$, then $M_K(x; \theta) \to J$ which is an attention matrix whose off-diagonal entries $J_{ij} = 1, i \neq j$, meaning deep features now equally contribute to others for refinement during training. The above procedure leads to our Kernel Attention Module (KAM) design as shown in Fig. 2, where its symbolic form can be summarized as:

$$x \leftarrow x + M_K(x; \theta)x = (I + M_K(x; \theta))x. \tag{1}$$

The proposed KAM mechanism can also be easily applied with multiple heads. We further note that in the implementation in Sect. 4, an extended form is used for $M_K^{ij} = \exp(-\alpha d(x_i, x_j)^2)$, where $\alpha \in (a, \infty)$. If the lower bound is set to $a = 0$, the case can be readily interpreted as $\alpha = 1/\sigma^2$ where the parameter can be understood as a kind of "standard deviation". Setting $a$ less than zero allows off-diagonal entries in $M_K$ to have values greater than one, in which case $a$ should be close to zero, i.e. $a \approx -\epsilon$ for $\epsilon$ small, to prevent numerical blow-up during training from poor matrix conditioning.

# 4   Experiments

In this research, we focused on model's performance on subject dependent classification tasks. In the spirit of [23] the data set was divided into non-overlapping epochs, each lasting one second, yielding $\sim$3300 epochs per trial, per subject, and $\sim$1060 epochs per labelled emotion. However, in contrast to [23], and most other studies where the training and testing data are manually split and models evaluated in a single pass, we adopt a cross-validation (CV) approach to improve evaluation robustness.

In the following benchmark using EEGNet with KAM, the data from each subject is split, taking 1/6 for validation during training. Five-fold cross validation (5-CV) is then performed on the remaining 5/6 of the data. Theses ratios are chosen to make the validation and test set roughly the same size during cross validation. For any model test, initial weights are set to the same values across each of the five folds. The network is trained for each fold over a maximum of 80 epochs, and the best model is selected at the epoch with the best validation accuracy. All experiments are trained with the same Adam optimizer configuration of an initial learning rate of $10^{-2}$ and a decay rate of 0.75, which only activates when no accuracy improvement is seen on the validation set in the past 10 epochs. A total of $5 \times 15 = 45$ training runs are conducted for each model compared in our benchmark. The code used for models' training and evaluation will be made available at https://github.com/dykuang/BCI-Attention.

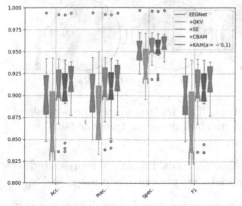

**Fig. 3.** Overall mean prediction performance across 15 subjects.

**Benchmark:** For benchmarking we compared five models: a) EEGNet, b) the basic $QKV$ type attention from Fig. 1, c) SE attention, d) CBAM attention, and e) KAM($a = -0.1$). All implementations herein are inserted at the same location shown in Fig. 2. Note that the basic QKV attention module does not perform well here, which is likely due to it, as mentioned in [6], being data thirsty and SEED not being a large enough dataset to quench. It is also worth mentioning here that the version of

KAM with $\alpha$'s lower bound $a = 0$ gives mean accuracy of 91.74% ± 3.02% which is slightly worse than the case of $a = -0.1$. This suggests that the extension of the lower bound to a negative value can potentially help during deep feature refinement. We also observe in our experiments that the training procedure for some subjects will push $\alpha$ slightly below zero for minimizing the loss function (see Fig. 5).

**Table 1.** Mean accuracy reported from different models.

| Models | EEGNet | +QKV | +SE | +CBAM | +KAM($a = -0.1$) |
|---|---|---|---|---|---|
| Parameters | 3851 | 4940 | 3933 | 4033 | 3852 |
| Acc(%) | 90.34 ± 3.69 | 86.81 ±4.19 | 91.20 ±3.42 | 90.40 ±3.97 | 91.89 ±2.76 |

**Channel Attention:** The inserted KAM module can potentially change the kernel weight originally designed in EEGNet during the model's decision process. Particularly, kernel weights in the first depthwise convolution layer (see Fig. 2) were treated as a representation for relative attention across different channels[1] in [12]. They can be affected when different attention modules are inserted. As the depthwise convolution applies one kernel to each of the eight channels, there are eight kernels associated to the architecture in Fig. 2. For clarity here, to illustrate the effect of these kernels on the network, we only examine the kernel applied on the first feature channel.

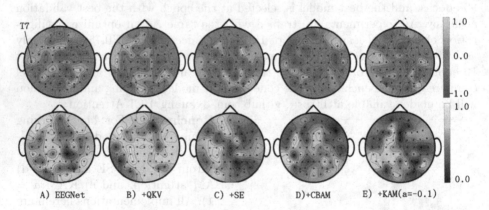

A) EEGNet      B) +QKV      C) +SE      D)+CBAM      E) +KAM(a=-0.1)

**Fig. 4.** Kernel weights mapped onto scalp maps. The first row shows the normalized mean. The second row shows the normalized standard deviation from the 5-CV.

To visualize the different spatial attention patterns discovered represented by the selected kernel weights of different models, we present scalp maps in Fig. 4.

---

[1] These are kernel weights in the first depthwise convolutional layer. The shape is of (1, 62) and can be directly associated with the 62 EEG sensor locations on scalp.

These maps are generated by training the models under 5-CV for subject S01, where mean and standard deviations are shown. It can be clearly seen overall that different modules can result in different channel attention patterns. While it may be difficult to immediately associate the mean value mappings to informative clinical interpretations, the spatial magnitude of the standard deviation does provide a way to measure different models' confidence in assigning kernel weights across different regions. For example, one thing to observe here is that all models visualized high mean attention values around the T7 region with relatively low uncertainty (represented by the std value). This observation seems to support some studies reporting correlations between emotional deficiency and memory development with specific temporal lobe function, such as in the diathesis of schizophrenia, e.g. [8], and in types of memory enhancement in forms of dementia, e.g. [11]. As an alternative research direction, how to enforce one's prior knowledge on task related scalp patterns so that the posterior learned patterns are robust to inserted attention modules is also important for building a better human interpretable model.

**Dependency on $\alpha$:** For better interpret the effect of module parameter $\alpha$ from KAM in trained models, we organised this section. Figure 5(A) shows the distribution over learned $\alpha$ during 5-CV for each subject. Among these, only experiments with data from subject S02 and S13 yield instances where $\alpha < 0$, while all other trainings find $\alpha \geq 0$. However, per subject speaking, the change in the lower bound on $\alpha$ does have a noticeable impact. For example, in the data from subject S05, S06, and S13 the learned $\alpha$ values cluster at locations close to zero, but we observed from our experiments that this small deviation from zero results in noticeable accuracy differences. This may because of the fact that small $\alpha$ will correspond to the case with large $\sigma$, i.e. further away from zero attention as explained above in Sect. 3.

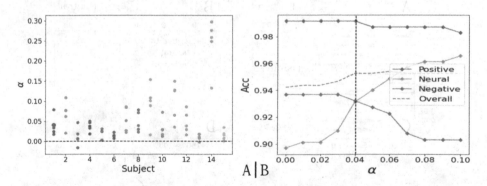

**Fig. 5.** *A*: Distribution of learned $\alpha$ value during the 5CV with EEGNet+KAM across the 15 subjects. *B*: Change of accuracy with varying value of $\alpha$ while freezing other parameters in the selected model (marked as red in first column i.e. subject S01 of *A*).

The one-parameter KAM design also makes it easy to analyze prediction accuracy as a function of the module parameter $\alpha$. As an example, we choose the model trained from the first fold (marked with a red dot in Fig.5(A)) for subject S01 and gather data from three trials each with a different emotion label for the dependency analysis. By varying $\alpha$ values in KAM while keeping other model parameters frozen, we can examine how $\alpha$ conditionally effects the prediction. In Fig. 5(B), it is interesting to see that the learned value (black vertical line) in KAM happens at a location where the overall accuracy line first rises to stabilize in this case. It is also interesting to observe that crossover between accuracy lines for "neutral" and "negative" happens at the same location of learned $\alpha$.[2]

This dependency can also be examined via the distribution of $\frac{\partial f_i(x)}{\partial \alpha}|_x$ for varying $\alpha$, where $x$ is the input and $f_i(x)$ is the output of the corresponding neuron (before activation) from the last dense layer for label $i \in \{1, 2, 3\}$, i.e. positive, neutral, and negative emotion labels respectively. The result is gathered in Fig. 6(A–C) computed with the same data mentioned in the previous paragraph. Of note, the partial dependencies appear to show very similar patterns. That is, the variance in each is a decreasing function for $\alpha \in [0, 0.1]$. This can be explained by the fact that $\alpha$ is packaged inside an exponential form that maintains its character through differentiation. Finally, Fig. 6(D) shows a close look at the histogram of how these distributions differ at the learned $\alpha = 0.0406$ from the model's selection during training.

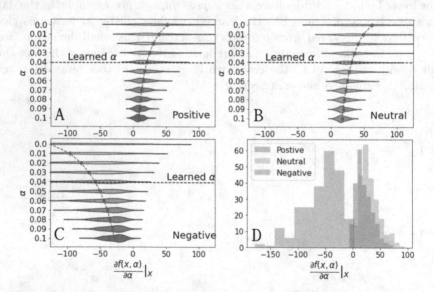

**Fig. 6. A–C:** Distribution at different $\alpha$ values corresponding to label "Positive", "Neutral" and "Negative". The mean is linked by dashed line. **D:** Distribution at learned $\alpha = 0.0406$ for different emotion labels.

---

[2] This might be an interesting coincidence since we also had other cases in our experiments where they do not meet exactly.

**Prediction Transition Curve:** In this section, we would like to explore how different models react when the input sample to be predicted is gradually transformed to another sample via some morphing operation $g$. In other words, we try to examine and interpret how different models react under a particularly selected "attack" $g$. Let $\{x_0, x_1, x_2\}$ denote 3 samples with labels 0, 1 and 2, and $\vec{p}_i = \{p_i^j\} = F(x_i)$ with $j = 0, 1, 2$ denoting model $F$'s confidence (here using a softmax score of the last dense layer) assigning input $x_i$ for label $j$. Then clearly by construction $\sum_j p_i^j = 1$ for any $i$. Further let $g_i^j(u), u \in [0, 1]$ denote an morphing operation between samples parameterized by $u$ such that $x_i = g_i^j(0) * x_i$ and $x_j = g_i^j(1) * x_i$. The symbol $*$ here stands for certain abstract action operation. In trinary classification tasks, as $u$ increases, $F[g_i^j(u)]$ will draw a curve in the hyperplane $x + y + z = 1$ inside the triangle formed by $[1, 0, 0], [0, 1, 0], [0, 0, 1]$. By checking these curves (we call them *prediction transition curves, PTC*), one can have an idea that how the trained model $F$ reacts with respect to the morphing operation $g$ on selected samples. The idea can be generated to higher dimensional cases with class categories $n > 3$, but it then becomes harder to visualize these cases as simple curves being embedded in higher dimensional simplices. This prediction transition curve provides a way for visualizing and interpreting model's predicting behavior under "attack" $g$ for given inputs.

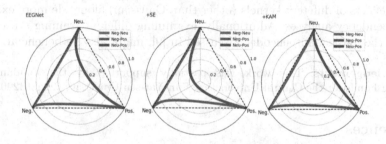

**Fig. 7.** The prediction transition curves from the three models on the same selected samples. From left to right: EEGNet, EEGNet+SE, EEGNet+KAM.

As a demonstration, we select three samples each with a different label, and EEGNet, EEGNet+SE, EEGNet+KAM all predict correctly on them. For simplicity, we choose the straightforward linear interpolation between samples for $g$, i.e. $g_i^j(u) * x_i = (1 - u)x_i + ux_j$. Notice that this definition of $g$ is symmetric in terms of $g_i^j(u) * x_i = g_j^i(1 - u) * x_j$. So morphing from $x_i$ to $x_j$ and $x_j$ from $x_i$ end up with the sample path (regardless of the direction). Other types of morphing operations can also be used depending on one's prior on the (known or inferred) underlying relationships between samples. The prediction transition curves obtained from morphing with $g$ are summarized in Fig. 7. It can be seen that all three compared models have almost straight PTCs for the connection of "Negative- Neutral" and "Negative-Positive", meaning that when one input sample is slowly morphing to the other, the model transits its confidence between the two labels almost linearly while leaving the third label almost untouched. Curves linking "Neutral-Positive" are all curved to the center at different extents, suggesting that models are in some sense "hesitating" to assign "Negative" for intermediate samples generated by morphing between 'Neutral" and "Positive". This example is consistent with the observations in Fig. 6 D) and provides a different angle suggesting that the trained models find it more difficult to separate "Positive" from "Neutral" emotions than separating "Negative" from "Positive" or "Neutral" from "Negative" emotions for the subject data under examination. An interesting followup question is whether this observation bears clinical significance as well, something which undoubtedly deserves consideration.

## 5  Conclusion

In this work, we present a kernel attention module that can be inserted into a network for deep feature refinement. Using EEGNet as the backbone model, the performance of KAM are benchmarked against several SOTA attention modules under cross validation with SEED dataset. With only one additional parameter, the idea behind KAM has demonstrated good potential for developing parameter efficient models that can simultaneously help human interpretation on trained models. Many follow-up studies are possible in this context, including investigating the effects of different kernels (other than Gaussian) alongside more exhaustive dependency analyses. Additionally, examining different training strategies, such as the masked-autoencoder discussed in [9], might also be beneficial.

**Acknowledgements.** This work was partially supported by the Fundamental Research Funds for the Central Universities, Sun Yat-sen University (22qntd2901).

## References

1. Bahdanau, D., Cho, K., Bengio, Y.: Neural machine translation by jointly learning to align and translate (2016)
2. Blankertz, B., et al.: Invariant common spatial patterns: alleviating nonstationarities in brain-computer interfacing. In: Advances in neural Information Processing Systems, pp. 113–120 (2008)
3. Blankertz, B., Lemm, S., Treder, M., Haufe, S., Müller, K.R.: Single-trial analysis and classification of ERP components–a tutorial. Neuroimage **56**(2), 814–825 (2011)
4. Cecotti, H., Graser, A.: Convolutional neural networks for p300 detection with application to brain-computer interfaces. IEEE Trans. Pattern Anal. Mach. Intell. **33**(3), 433–445 (2010)
5. Congedo, M., Barachant, A., Bhatia, R.: Riemannian geometry for EEG-based brain-computer interfaces; a primer and a review. Brain-Comput. Interf. **4**(3), 155–174 (2017)
6. Dosovitskiy, A., et al.: An image is worth 16x16 words: Transformers for image recognition at scale. arXiv preprint arXiv:2010.11929 (2020)

7. Fazli, S., Popescu, F., Danóczy, M., Blankertz, B., Müller, K.R., Grozea, C.: Subject-independent mental state classification in single trials. Neural Netw. **22**(9), 1305–1312 (2009)
8. Goghari, V.M., MacDonald, A.W., III., Sponheim, S.R.: Temporal lobe structures and facial emotion recognition in schizophrenia patients and nonpsychotic relatives. Schizophr. Bull. **37**(6), 1281–1294 (2011)
9. He, K., Chen, X., Xie, S., Li, Y., Dollár, P., Girshick, R.: Masked autoencoders are scalable vision learners. arXiv preprint arXiv:2111.06377 (2021)
10. Hu, J., Shen, L., Sun, G.: Squeeze-and-excitation networks. In: Proceedings of the IEEE Conference on Computer Vision and Pattern Recognition, pp. 7132–7141 (2018)
11. Kumfor, F., Irish, M., Hodges, J.R., Piguet, O.: Frontal and temporal lobe contributions to emotional enhancement of memory in behavioral-variant frontotemporal dementia and Alzheimer's disease. Front. Behav. Neurosci. **8**, 225 (2014)
12. Lawhern, V.J., Solon, A.J., Waytowich, N.R., Gordon, S.M., Hung, C.P., Lance, B.J.: EEGNet: a compact convolutional neural network for EEG-based brain-computer interfaces. J. Neural Eng. **15**(5), 056013 (2018)
13. Li, J., Zhang, L.: Bilateral adaptation and neurofeedback for brain computer interface system. J. Neurosci. Methods **193**(2), 373–379 (2010)
14. Liu, G., Huang, G., Meng, J., Zhang, D., Zhu, X.: Improved GMM with parameter initialization for unsupervised adaptation of brain-computer interface. Int. J. Num. Methods Biomed. Eng. **26**(6), 681–691 (2010)
15. Liu, G., Zhang, D., Meng, J., Huang, G., Zhu, X.: Unsupervised adaptation of electroencephalogram signal processing based on fuzzy c-means algorithm. Int. J. Adapt. Control Signal Process. **26**(6), 482–495 (2012)
16. Liu, Z., et al.: Swin transformer: hierarchical vision transformer using shifted windows. In: Proceedings of the IEEE/CVF International Conference on Computer Vision, pp. 10012–10022 (2021)
17. Lotte, F., Bougrain, L., Cichocki, A., Clerc, M., Congedo, M., Rakotomamonjy, A., Yger, F.: A review of classification algorithms for EEG-based brain-computer interfaces: a 10 year update. J. Neural Eng. **15**(3), 031005 (2018)
18. Lotte, F., Congedo, M., Lécuyer, A., Lamarche, F., Arnaldi, B.: A review of classification algorithms for EEG-based brain-computer interfaces. J. Neural Eng. **4**(2), R1 (2007)
19. Lu, N., Li, T., Ren, X., Miao, H.: A deep learning scheme for motor imagery classification based on restricted Boltzmann machines. IEEE Trans. Neural Syst. Rehabil. Eng. **25**(6), 566–576 (2016)
20. Schlögl, A., Vidaurre, C., Müller, K.R.: Adaptive methods in BCI research-an introductory tutorial. In: Graimann, B., Pfurtscheller, G., Allison, B. (eds.) Brain-Computer Interfaces, pp. 331–355. Springer, Heidelberg (2009). https://doi.org/10.1007/978-3-642-02091-9_18
21. Steyrl, D., Scherer, R., Faller, J., Müller-Putz, G.R.: Random forests in non-invasive sensorimotor rhythm brain-computer interfaces: a practical and convenient non-linear classifier. Biomed. Eng. Biomedizinische Technik **61**(1), 77–86 (2016)
22. Woo, S., Park, J., Lee, J.Y., Kweon, I.S.: CBAM: convolutional block attention module. In: Proceedings of the European Conference on Computer Vision (ECCV), pp. 3–19 (2018)
23. Zheng, W.L., Lu, B.L.: Investigating critical frequency bands and channels for EEG-based emotion recognition with deep neural networks. IEEE Trans. Auton. Ment. Dev. **7**(3), 162–175 (2015)

# Explainable Artificial Intelligence for Breast Tumour Classification: Helpful or Harmful

Amy Rafferty[1]($\boxtimes$) , Rudolf Nenutil[2], and Ajitha Rajan[1]

[1] University of Edinburgh, 10 Crichton Street, Edinburgh EH8 9AB, UK
a.rafferty@live.com
[2] Masaryk Memorial Cancer Institute, Žlutý kopec 543/7,
602 00 Brno-střed-Staré Brno, Czechia

**Abstract.** Explainable Artificial Intelligence (XAI) is the field of AI dedicated to promoting trust in machine learning models by helping us to understand how they make their decisions. For example, image explanations show us which pixels or segments were deemed most important by a model for a particular classification decision. This research focuses on image explanations generated by LIME, RISE and SHAP for a model which classifies breast mammograms as either benign or malignant. We assess these XAI techniques based on (1) the extent to which they agree with each other, as decided by One-Way ANOVA, Kendall's Tau and RBO statistical tests, and (2) their agreement with the diagnostically important areas as identified by a radiologist on a small subset of mammograms. The main contribution of this research is the discovery that the 3 techniques consistently disagree both with each other and with the medical truth. We argue that using these off-shelf techniques in a medical context is not a feasible approach, and discuss possible causes of this problem, as well as some potential solutions.

**Keywords:** Machine learning · Breast tumour classification · Explainable AI · LIME · RISE · SHAP

## 1 Introduction

Recent developments in deep learning (DL) have sparked an interest in more high-stakes applications such as medical diagnostics. Given a medical scan, a clinician may want to differentiate between healthy and unhealthy tissue, or between pathologies. However, the black-box nature of DL models means their conclusions tend not to be trusted by clinicians who cannot determine how the model came to its decision. Interpretable explanations are therefore crucial. Many medical experts have already expressed their concerns over rising black-box DL approaches [8].

Explainable AI (XAI) techniques exist to bridge this gap by intuitively highlighting the most important features of an input. This gives the model practitioner more information about how to improve the model's correctness, and gives the end-user, potentially a non-expert, an idea of how the model came to its conclusion. Knowing that a model's conclusion is correct is essential in medical diagnostics as their outcomes could impact lives.

M. Reyes et al. (Eds.): iMIMIC 2022, LNCS 13611, pp. 104–123, 2022.
https://doi.org/10.1007/978-3-031-17976-1_10

Using XAI techniques for medical diagnostics comes with its own set of problems. Medical datasets are problematic due to differing labelling standards - some images have complete clinical annotation, while others simply state whether a tumour is present. Many techniques run into problems for images with small regions of interest (ROIs), due to their usage of image segmentation. This is the case for breast mammograms as cancerous regions can be extremely small. [23] discusses the serious implications of bad explanations in high stakes contexts. Saliency maps, which are commonly used to visualise image explanations, can be virtually identical for different classes on the same image [2]. Unreliable and misleading explanations can have serious negative implications.

We present a case study which focuses on the quality of explanations from 3 widely used XAI techniques, applied to a publicly available CNN-based classification model used to identify malignant and benign breast tumours (originally designed for brain tumour detection [15]), and a public anonymised dataset of benign and malignant breast mammograms [12]. We assess the XAI techniques based on (1) the extent to which they agree with each other for the whole dataset, and (2) evaluation by two independent radiologists on the correctness of the important regions identified by each of the XAI techniques for 10 mammograms. The XAI techniques used in our study, LIME [20], SHAP [14] and RiSE [18], are discussed in the next Section.

## 2    Related Work

Many existing XAI techniques are applicable to the medical context. [31] presents an exhaustive list of techniques used for medical image analysis - we limit our consideration here to LIME, SHAP and RISE due to their popularity and ease of use [5]. We plan to consider other XAI techniques in the future.

**LIME - Local Interpretable Model-Agnostic Explanations.** LIME [20] is an XAI technique which can be applied to any model without needing any information about its structure. LIME provides a local explanation by replacing a complex neural network (NN) locally with something simpler, for example a Ridge regression model. LIME creates many perturbations of the original image by masking out random segments, and then weights these perturbations by their 'closeness' to the original image to ensure that drastic perturbations have little impact. It then uses the simpler model to learn the mapping between the perturbations and any change in output label. This process allows LIME to determine which segments are most important to the classification decision - these segments are then shown in the visual explanation output.

**RISE - Randomized Input Sampling for Explanations of Black Box Models.** RISE [18] works by first generating many random masks of an image, multiplying them elementwise with the image, and then feeding them directly into the original model for label prediction. Saliency maps are generated from a linear combination of the masks where weights come from the output probabilities predicted by the model. These saliency maps highlight the most important pixels of the image regarding its classification. This makes RISE extremely interpretable. RISE is also model agnostic. We note that RISE is very similar to LIME, however it measures saliency based on individual pixels, rather than superpixels, and therefore may perform better on images with small ROIs (e.g. mammograms).

**SHAP - Shapley Additive Explanations.** SHAP [14] is another model-agnostic approach which uses Shapley values, a concept from game theory, to find the contribution of each feature to the model's output. The image is segmented to reduce the number of value computations. Starting from one random segment, we add one segment at a time until the correct model classification is possible. This is repeated many times with random orderings to get the importance of each segment, represented as Shapley values. Large positive SHAP values indicate that the segment is very important to the classification decision. SHAP is also a highly interpretable technique. We note that SHAP values are derived from game theory's Shapley values - they are not the same, and the mathematical differences are discussed in detail in [14].

## 2.1 XAI in Medicine

These methods, as well as other techniques [24, 26, 27, 29, 30, 33, 34], have had huge success, particularly in the image classification and Natural Language Processing fields, however they are only beginning to be evaluated in any medical context [31]. An important issue to note is that when using larger medical images such as MRI scans, there is a need to split the images into tiles due to their extremely high resolutions. XAI techniques then need to be run on each tile, and the results need to be brought back together. Since we are working with mammograms, this is not an issue for this research, but is something we plan to explore in future work.

[25] highlights some of the challenges faced by medical professionals regarding XAI - not all visualisations are interpretable, there is no current definition for sufficient explainability in the field, and XAI techniques are not satisfactorily robust [1]. They describe the issue of the knowledge gap between AI and medical professionals, and the effects this has on techniques. Currently the focus of medical XAI seems to be on diagnosing rare diseases and monitoring health trends [25]. Some contributions to XAI for tumour classification exist, for example [6] which focuses on sequencing gene data, and [21] which also focuses on mammograms, though with gradient-based XAI techniques. [11] argues that explanations generated by LIME and SHAP cause no improvement on human decision making abilities - when shown an image both with and without an explanation, there was no statistical difference in the time it took for people to classify the image by eye, or in the number of mistakes made. This is concerning as the goal of diagnostic XAI is to make the lives of medical professionals easier, and remove the need for tedious by-eye classification [10, 19].

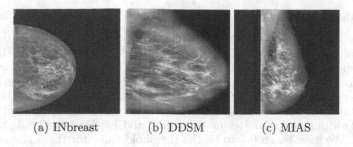

(a) INbreast          (b) DDSM          (c) MIAS

**Fig. 1.** Example images from each of the three dataset sources.

## 3    Model Setup

### 3.1    Data Pre-Processing

For this research we take breast mammograms with cancerous masses from a public dataset [12], which takes images from 3 official datasets - INbreast [16], DDSM [3] and MIAS [28]. When generating this dataset, the creators [13] extracted a small number of images with masses from each source, and performed data augmentation in the form of image rotation to generate a larger dataset. They also re-sized images to $227 \times 227$ pixels.

The public dataset [12] we are using is large. The original paper introducing this dataset [13] details their data augmentation techniques, which includes rotating and flipping each image to generate 14 variations of itself. This is not useful for this research - we are not trying to train a model that can cope with rotated breast scans, as the original scans and therefore any unseen real-world scans are all of the same orientation. We only use images of the original orientation. We also only take images from INbreast and DDSM, as the only MIAS scans present in the dataset were benign, though we plan to include MIAS for evaluation purposes (e.g. Out-Of-Distribution detection (OOD) [17]) in future work to improve model confidence. The visual difference in original scans between the 3 sources is shown in Fig. 1. After selecting the images of the same orientation from the INbreast and DDSM sections of the dataset, we have a dataset of 2236 images - 1193 benign and 1043 malignant.

**Image Cropping.** For maximal model performance, we crop out as much of the black background as possible, making the breast the focus of each image. This was performed using basic Python opencv code. Images are then resized to the original $227 \times 227$ pixel format for consistency.

**Dataset Split.** Our dataset of 2236 images is split into a (Training/Validation/Testing) ratio of (2124/56/56). The Validation set will be used for all intermediate experiments - deciding how many epochs to train the model for, and tuning parameters for LIME. A small test set was chosen to ensure sufficient model training due to the small dataset size.

## 3.2    Model Architecture

We use an existing public CNN [15] which was originally used for binary classification of brain scans regarding the presence of a tumour. We use this model as it was specifically designed for the domain of tumours in medical scans, and was therefore reliable in the sense that it was likely to perform well on data like ours - noisy black and white scans containing cancerous legions. In the original study the model achieved 88.7% accuracy on the test set. The model takes an image and outputs a decimal value between 0 and 1, where 0 is benign, and 1 is malignant. We have taken 0.5 to be the threshold value for these classifications. The CNN contains 8 layers, using ReLU activation.

To avoid overfitting, we train four models differing only in numbers of epochs, and evaluate their performances on the Validation set. The performances of these models are described in Table 1 (Appendix A.1) in the form of their Accuracy and F1 Score. These statistics are based on a Validation set of 56 images. We proceed with the 75 epoch model as it has the highest performance scores. On the Test set, the 75 epoch model has an Accuracy of 0.9643 and an F1 Score of 0.9642. For training, we use Keras with the adam optimizer and binary cross-entropy loss function. We use a Intel(R) Core(TM) i5-7300HQ CPU @ 2.50GHz processor laptop for our experiments.

Although the accuracy of the model on the test set is high (96.43%), it is not clear whether the model infers classifications using the correct image features. In medical diagnostics, it is imperative that a clinician is able to interpret and understand the reasons for the classification label. Explanations from XAI techniques are meant to address this need.

## 4    Explanations

We generated individual explanations using each of the 3 XAI techniques, for a test set of 56 images. For illustration, we show explanations for the same 6 benign and 6 malignant examples with each XAI technique in Appendix A.8 (LIME Fig. 6, RISE Fig. 7, SHAP Fig. 8). Code associated with generating explanations can be found at https://anonymous.4open.science/r/EvaluatingXAI-11DF/.

### 4.1    LIME

Our Python code for generating LIME explanations follows the steps described by [20]. For image segmentation, we used Python's scikit-image quickshift algorithm with empirically chosen parameters. When generating explanations, we highlight the boundaries of the L most important features for visibility. L was empirically chosen.

**Choosing Segmentation Parameters.** For segmentation we use the scikit-image quickshift algorithm, which has 3 tuneable parameters - kernel size, max-dist, and ratio. These parameters and their effects are detailed in their documentation [7]. We use a small kernel size of 2, a default max-dist value of 10, and a small ratio value of 0.1. This was because we wanted many small segments with little emphasis on colour boundaries, to ensure that we consider small regions of interest (ROIs), and do not quantify the pixels at the boundary of the breast as incorrectly important.

**Choosing L.** We define L as the number of most important features used in LIME explanations. The L value will determine the features shown in our LIME explanations, and also how many pixels are compared in the later One-Way ANOVA analysis. We will be comparing lists of most important pixels as decided by each XAI technique - the lengths of these lists will be equal for the 3 techniques, and will be the number of pixels within the L most important LIME features for a given image. We will then calculate the % pixel agreement between each pair of methods, defined as the proportion of pixels the lists have in common. To determine our L value, we calculate the average % pixel agreements between methods using L values of 3, 4, 5, 6 and 7. Averages are taken over the first 30 images in the Validation set for the sake of time. The results are shown in Fig. 4 (Appendix A.2). As L increases, average pixel agreement increases between each pairwise technique comparison. It is infeasible to keep increasing L as we are trying to compare only the most important pixels. We have chosen L to be 6 as the first decrease in average agreement between all three techniques occurs at L = 7. Also, in the case of the pairwise comparisons LIME-SHAP and LIME-RISE, the jump in agreement from 6 to 7 is much smaller than from 5 to 6.

**Observations.** Figure 6 (Appendix A.8) shows 12 examples of LIME explanations - 6 for benign scans and 6 for malignant. For both classes, some explanations highlight undesirable features such as the image background. This is likely due to the variance in breast shape throughout the dataset, which can clearly be seen in these examples. This effect could be reduced by using larger datasets in the future. Looking at these explanations without ground truth tells us little about whether they are highlighting genuine cancerous regions. To evaluate LIME's performance, we compare its outputs to those of RISE and SHAP, and to a radiologist's evaluation.

### 4.2  RISE

Our Python code for RISE follows the steps described by [18]. Figure 7 (Appendix A.8) shows 12 examples of RISE explanations - 6 for benign scans and 6 for malignant. In these heatmaps, the most important pixels are shown as red, and the least important are shown as blue. We note that images have different importance value scales.

**Observations.** RISE generally assigns background pixels a medium relative importance. We expect that this is again due to irregular breast shapes. LIME and RISE seem to generate poor results for the same images - we define poor results as explanations which highlight background regions as important. Figure 7(j) shows a case where RISE performs poorly for a malignant scan. LIME also performs poorly on this image, shown in Fig. 6(j). This image has an irregular shape, which supports our thoughts. Figure 7(c) and Fig. 6(c) show the same issue for a benign scan.

### 4.3  SHAP

Our SHAP explanation code follows the steps described by [14]. Default values were used for image segmentation, and SHAP's Kernel Explainer was used.

Figure 8 (Appendix A.8) shows 12 examples of SHAP explanations - 6 for benign scans and 6 for malignant. Segments that contribute the most to the classification of the image are shown as green. The least important segments are shown as red. We note that SHAP value scales are not consistent across images.

**Observations.** Figure 8(j) shows that SHAP performs poorly for the 4th malignant scan, much like LIME and RISE - heavily influential segments exist at the top-left corner of the image, which are background pixels. In most cases, the superpixels outside the boundary of the breast seem to have low SHAP values. SHAP seems to generally disregard background pixels with more success than RISE.

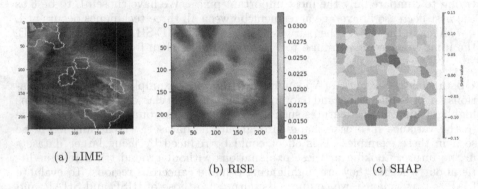

(a) LIME           (b) RISE           (c) SHAP

**Fig. 2.** Explanations by LIME, RISE and SHAP for a benign mammogram.

## 5 Evaluating Explanations

Looking at the 3 explanations side-by-side for an image, as in Fig. 2, we can start to infer some agreement. However, due to the different explanation formats between techniques, the amount of agreement is unclear. In addition to visualisations, we use statistical analysis to compare the importance rankings of pixels between XAI techniques.

**Visualising Agreement.** We use the 6 most important features in our LIME explanations, and denote n to be the number of pixels within these features. We visualise overlap between the n most important pixels given by each XAI technique, as in Fig. 3. Generally, there are always areas which all 3 techniques identify as highly important. However there are more regions where they disagree. Sub-figures (b) and (d) from Fig. 3 show cases where explanations have performed poorly - defined as identifying background pixels as most important. This is likely due to irregular breast shapes within the dataset. Figures 3(a) and (c) show cases where explanations have multiple clear points of agreement.

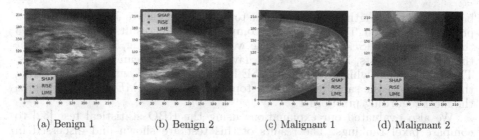

(a) Benign 1      (b) Benign 2      (c) Malignant 1      (d) Malignant 2

**Fig. 3.** Overlap between the 3 XAI techniques regarding the n most important pixels. SHAP is blue, RISE is red, LIME is green. (Color figure online)

## 5.1 One-Way ANOVA

One-Way ANOVA [22] compares the means of two or more groups for a dependent variable. Our input groups are 3 lists of % pixel agreements between methods for each test set image, labelled LIME-RISE, LIME-SHAP, and RISE-SHAP. To generate these lists, we identify the n most important pixels according to each technique, where n is the number of pixels within the top 6 LIME features for a given image. This is because LIME outputs binary values for each pixel (presence in the L most important features) while RISE and SHAP assign decimal importance values. We then calculate the % pixel agreement across each pair of pixel lists for each image. We define % pixel agreement as the proportion of pixels the lists have in common. Results are in Table 2 in Appendix A.3.

The only statistically significant test is Test 2, shown in Table 2. This tells us that of all pairwise comparisons, there is only statistically significant difference in average pixel agreement between the comparisons of LIME-RISE and RISE-SHAP. Analysing the statistical composition of the pixel agreement lists supports this conclusion. Figure 5 (Appendix A.4) visualises these results. The largest difference in mean (green triangles) is between LIME-RISE and RISE-SHAP. Figure 5 and Table 3 (in Appendix A.4) show that the average pixel agreement between techniques is startlingly low - 20–30% on pairwise comparisons, and under 10% when comparing all three. However, these values still represent significant numbers of pixels, as our images are large and have small ROIs.

## 5.2 Kendall's Tau

Kendall's Tau [9] is a measure of the degree of correlation between two ranked lists. The purpose of Kendall's Tau is to discover whether two ordered lists are independent. We perform this test using the built-in Python scipy method, and set the inputs to be the ordered lists of pixels and their importance values for each of the 3 XAI techniques, in the form "(x, y): value".

We apply Kendall's Tau to each test set image 3 times - on the full pixel list, on the top n most important pixels, and on the top 1000. We want to discover any statistically significant correlation regarding the most important pixels to the classification between techniques - if there is, and the Tau values are positive, this implies agreement. Results are shown in Table 5 (in Appendix A.6). We use an alpha value of 0.05. From Table 5 we can conclude that the only instances of statistically significant correlation come from the LIME-SHAP comparison -

both on the full length pixel list and the top n pixels. Positive Tau values imply a positive correlation. The LIME-RISE comparison yields results closer to the 0.05 threshold while RISE-SHAP yields the worst results. In Fig. 5, we saw that for the top n pixel lists, LIME and RISE have the highest mean pixel agreements. This implies that while LIME and RISE have higher pixel agreement regarding the presence of the same pixels in the top n pixel lists, LIME and SHAP agree the most regarding pixel order.

We also evaluated our explanations using the RBO statistical test [32] to compare pixel rankings. The results of this test are shown and discussed in Appendix A.5.

## 5.3   Radiologist Evaluation

To assess our explanations with respect to the medical truth as understood by a clinician, we consulted 2 independent radiologists and provided them with a subset of 10 images - 5 with benign and 5 with malignant classification, each of them associated with explanations from the 3 different techniques. We were unable to gather an expert evaluation for the entire test set due to limited availability of the radiologists, though we intend to expand this form of XAI technique evaluation in future work.

The results gathered from this evaluation with 10 images for the first and second radiologists are shown in Tables 6 and 7 (in Appendix A.7). In these tables,'B' in the column heading refers to a benign image and 'M' refers to a malignant image. We requested the radiologists to score each explanation between 0 and 3 to represent its agreement to radiologist identified image regions. The definition of the scores provided to the radiologists are as follows:

**0** = Explanation completely differs from expert opinion
**1** = Explanation has some similarities, but mostly differs from expert opinion
**2** = Explanation mostly agrees with expert opinion, though some areas differ
**3** = Explanation and expert opinion completely agree

It is worth noting that no explanation earned a label of 3 from either radiologist - each explanation either identified erroneous regions or missed important sections. LIME appears to perform the worst within this subset of 10 images, while RISE performs the best. This is likely because RISE is the only method which uses pixels rather than superpixels and is therefore more fine grained when examining image regions and less likely to miss small regions of interest. There does not seem to be any difference in explanation quality between benign and malignant images for any technique.

The radiologists noted the following limitations with the explanation techniques:

- None of the explanations could identify the entire tumour region. Explanation methods only highlight fragmented relevant regions and this is along with many irrelevant regions.
- Explanations for both malignant and benign tumours are distributed all over the image and fail to take into account clinical features, like shapes of masses, margins, the density of tissues, and structural distortion.

Our radiologist evaluation using 10 mammograms may not be representative of a real world dataset. However, the issues highlighted by these comments are consistent problems - this will be discussed in Sect. 6.

## 5.4    Threats to Validity

This research uses a small dataset of breast mammograms which may not be representative of the population. We limit our classification task to benign or malignant - in reality there are many types of lesion for both classes, which would appear differently in mammograms. In future work, using a non-binary classifier alongside a more thorough radiologist evaluation may allow us to better analyse the failures of our techniques. We have assumed that the low cohesion between XAI techniques paired with the high model test accuracy indicates that failures are due to the XAI techniques, and not the model itself. In future work we will utilise multiple models and explore alternate XAI evaluation techniques [4] in order to back up this claim. All empirical analysis regarding LIME parameter tuning and the choice of L was based solely on the patterns within our data. They may not hold up when compared to a larger dataset. Our XAI techniques by definition utilise randomization when generating masks, therefore re-running our code will generate slightly different results to the ones displayed here. This variation is not hugely impactful as we generally discuss average values in our statistical tests. Our code for LIME, RISE and SHAP is not the only way of implementing these techniques - there are many public examples which implement the steps described in the literature in slightly different ways. Because of this, another researcher's code may yield different results to the ones shown here.

# 6    Observations and Discussion

**Each Technique Performs Poorly on the Same Images.** Our explanations highlight the quality variation within the test set. Each XAI technique performed poorly (highlighted background pixels as most important) on the same images, usually mammograms with irregular breast shapes. This is likely due to our small dataset and the effect of blurring and image re-sizing. It's interesting to note that these problems don't seem to impede the model accuracy, only the quality of explanations.

**Percentage Pixel Agreement Between XAI Techniques is Extremely Low.** LIME and RISE appear to have the most pixel agreement according to One-Way ANOVA. However, these values are not high, with an average agreement of 28%. Combining Kendall's Tau with One-Way ANOVA, we find that while LIME and RISE consistently highlight the highest proportion of the same important pixels, LIME and SHAP have the most similar pixel orderings. This is supported by RBO.

**Radiologist Evaluation Revealed Explanations from All Three Techniques Were Unhelpful.** The radiologists found that RISE performed marginally better than the other two techniques. Explanations from all three techniques, however, do not consider clinical features within mammograms that are used to diagnose benign or malignant tumours, such as shape of mass, boundary, and density. The explanations do not highlight the entire tumour as important, but instead sparsely pick parts of the tumour along with many irrelevant regions. The XAI techniques we have used have low levels of agreement with each other, as well as low levels of agreement with the medical truth.

## 6.1   Discussion

The goal of this research was to determine whether taking off-shelf XAI techniques and applying them to breast tumour classification was a feasible approach that would hold up in the real world. Bringing together our observations tells us that this is not the case.

Though LIME and SHAP have the highest agreement in pixel orderings, these agreement levels are still very low. Explanations from these techniques highlight some common areas, though have significant disagreements and are therefore unreliable for use in diagnostics. The most likely reason that LIME and SHAP have the highest pixel ordering agreement is that these methods both utilise superpixels, while RISE does not. Discussing similarities in pixel orderings is problematic in this context, due to the differing ways in which each of the 3 XAI techniques assign importance values to pixels. We note that these differences come from both the underlying properties of each technique, and from our code architecture. LIME's binary scoring method is likely the reason behind the slightly higher % pixel agreement statistics for pairwise comparisons involving LIME.

Each XAI technique works differently, and resulting explanations depend on many different factors - segmentation, mask randomization, and tuneable parameters. While this is an expected reason for some result variation, a higher level of cohesion in explanations was to be expected. We identified that each technique incorrectly highlighted background regions as being most important on images with irregular breast shapes. While this may have been caused by the small size of the dataset, and image quality after pre-processing, we would have expected the model's accuracy to also decline to reflect this, and it did not. We also note that the techniques showed no difference in explanation quality for images from the benign or malignant classes.

Regarding the medical truth according to a radiologist, RISE seems to produce the most medically correct explanations, while the results of LIME and SHAP are often entirely incorrect. This is likely because RISE involves no image segmentation. No explanations were labelled as perfect - areas are always missed or incorrectly highlighted. We therefore conclude that explanations generated by LIME, RISE and SHAP are in disagreement with respect to both each other, and to the medical truth, and so do not perform reliably in this context. The results of these explanation techniques do not match or consider what a radiologist would want in a real-world context. Instead of pixels or superpixels, techniques should identify clinically defined regions. This is a gap that needs to be bridged - we highlight the need for specific, carefully defined techniques for explaining tumour images that take clinical features into account.

## A   Appendix

### A.1   Model Training Results

The results of the experiment used to choose the 75 epoch model when considering the impact of overfitting on our CNN, as discussed in Sect. 3.2 of this report.

**Table 1.** Validation accuracy and F1 Score for CNNs trained with different numbers of epochs.

| Epochs | Accuracy | F1 score |
|--------|----------|----------|
| 25     | 0.8214   | 0.7917   |
| 50     | 0.8214   | 0.7917   |
| **75** | **0.8750** | **0.8571** |
| 100    | 0.8036   | 0.7660   |

## A.2   Choosing L Parameter for LIME

The results of the experiment used to choose L, as discussed in Sect. 4.1 of this report.

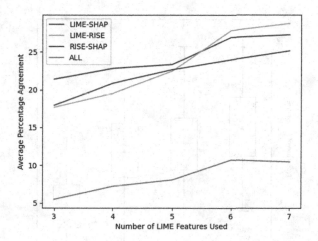

**Fig. 4.** Average % pixel agreement values between techniques taken over 30 images from the Validation set.

## A.3   One-Way ANOVA Results

We present here the statistical hypotheses used for the One-Way ANOVA test, as well as the results gathered. This statistical test and its implications is discussed in Sect. 5.1 of this report. The results are shown in Table 2.

The hypotheses for One-Way ANOVA are as follows:

- H0: There is no statistically significant difference between the means of the groups.
- H1: There is a statistically significant difference between the means of the groups.

**Table 2.** Results of One-Way ANOVA tests as described in the text. **Bold** results are statistically significant (alpha value 0.05).

| Test | Methods compared | F-statistic | p-value |
|------|------------------|-------------|---------|
| 1 | LIME-RISE, LIME-SHAP | 3.7823 | 0.0544 |
| 2 | LIME-RISE, RISE-SHAP | **9.1855** | **0.0031** |
| 3 | RISE-SHAP, LIME-SHAP | 1.6193 | 0.2060 |

### A.4    Pixel Agreement Statistics

Figure 5 presents a box plot representation of the % pixel agreement values between XAI techniques, taken over all images in our test set. These results are discussed in Sect. 5.1 of this report. Table 3 also represents these agreement values.

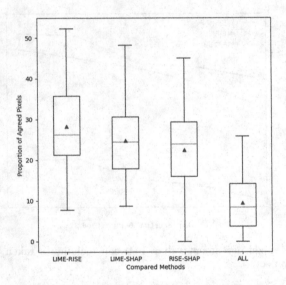

**Fig. 5.** % Pixel Agreement between techniques for n most important pixels. Medians are orange lines, means are green triangles. (Color figure online)

### A.5    Ranked Biased Overlap (RBO) Results

RBO [32] weights each rank position by considering the depth of the ranking being examined, minimising the effect of the least important pixels. Taking two ranked lists as inputs, RBO outputs a value between 0 and 1, where 0 indicates that the lists are disjoint, and 1 indicates that they are identical. The results of RBO depend on the tuneable parameter p [32]. Small p values place more weight on items at the top of an ordered list. While this is desirable, we must consider the difference in pixel importance value allocation methods between techniques. RISE applies a decimal score to each pixel. SHAP applies the same decimal score to each pixel within a given image segment. LIME uses binary values indicating

**Table 3.** Statistical overview of percentage pixel agreements for all method comparisons.

| Techniques | Mean | Std | Min | Max |
|---|---|---|---|---|
| LIME-RISE | 28.27% | 10.13% | 7.74% | 52.19% |
| LIME-SHAP | 24.73% | 8.75% | 8.73% | 48.16% |
| RISE-SHAP | 22.45% | 9.82% | 0.00% | 44.97% |
| ALL | 9.48% | 6.18% | 0.00% | 25.88% |

whether the pixels are in the top 6 most important features. We use large p values to properly encompass similarities between larger groups of pixels with identical values.

Table 4 shows the average, minimum and maximum RBO values for each pairwise pixel list comparison. The average RBO values for each comparison tell us that the pixel lists are almost disjoint. This is expected due to the differing pixel importance allocation methods as discussed. Instead we consider the maximum values - LIME and SHAP generate lists that are hugely identical for at least one instance in the test set, with maximum RBO values in the range 0.69–0.78. The other pairwise comparisons do not come close to these numbers. This observation supports Kendall's Tau - both tests have identified LIME and SHAP as the techniques with the highest agreement regarding pixel orderings.

**Table 4.** RBO results performed on full ordered pixel importance lists for each technique, with differing p values. Values shown to 3 decimal places, though we note here that these values are never exactly zero, just extremely small.

| - | RISE-SHAP | | | LIME-SHAP | | | LIME-RISE | | |
|---|---|---|---|---|---|---|---|---|---|
| p | 0.9 | 0.95 | 0.99 | 0.9 | 0.95 | 0.99 | 0.9 | 0.95 | 0.99 |
| Min | 0.000 | 0.000 | 0.000 | 0.000 | 0.000 | 0.000 | 0.000 | 0.000 | 0.000 |
| Max | 0.000 | 0.000 | 0.077 | 0.697 | 0.763 | 0.782 | 0.002 | 0.023 | 0.265 |
| Avg | 0.000 | 0.000 | 0.003 | 0.019 | 0.027 | 0.045 | 0.000 | 0.001 | 0.011 |

## A.6   Kendall's Tau Results

We present here the statistical hypotheses used for the Kendall's Tau test, as well as the results gathered. This statistical test and its implications is discussed in Sect. 5.2 of this report. The results are shown in Table 5.
The following hypotheses are used:

– H0: There is no statistically significant correlation, the lists are independent.
– H1: There is a statistically significant correlation in pixel orderings between lists, they are not independent.

**Table 5.** Kendall's Tau comparison results. n is defined in the text. Values are averages taken over the test set, shown to 3 decimal places. **Bold** results are statistically significant.

| Techniques | p-values | | | Tau | | |
|---|---|---|---|---|---|---|
| | Full | n | 1000 | Full | n | 1000 |
| RISE-SHAP | 0.123 | 0.125 | 0.249 | 0.003 | 0.002 | 0.001 |
| LIME-SHAP | **0.000** | **0.048** | 0.067 | **0.154** | **0.106** | 0.293 |
| LIME-RISE | 0.066 | 0.055 | 0.133 | 0.004 | −0.006 | 0.014 |

## A.7 Radiologist Opinions

Here we present the results as received from two independent radiologists, as well as definitions of the scoring system used to evaluate explanations,

We requested each explanation be scored between 0 and 3 to represent its agreement to radiologist identified image regions. The definition of the scores provided to the radiologists are as follows:

0 = Explanation completely differs from expert opinion
1 = Explanation has some similarities, but mostly differs from expert opinion
2 = Explanation mostly agrees with expert opinion, though some areas differ
3 = Explanation and expert opinion completely agree

**Table 6.** Radiologist evaluation regarding explanations generated on a subset of 10 images. B denotes benign, and M denotes malignant.

| Image | B1 | B2 | B3 | B4 | B5 | M1 | M2 | M3 | M4 | M5 |
|---|---|---|---|---|---|---|---|---|---|---|
| LIME | 0 | 1 | 0 | 1 | 1 | 0 | 0 | 1 | 0 | 0 |
| RISE | 0 | 1 | 1 | 1 | 1 | 2 | 1 | 1 | 1 | 2 |
| SHAP | 0 | 0 | 0 | 1 | 2 | 0 | 1 | 1 | 1 | 0 |

**Table 7.** Second radiologist evaluation regarding explanations generated on a subset of 10 images. B denotes benign, M denotes malignant.

| Image | B1 | B2 | B3 | B4 | B5 | M1 | M2 | M3 | M4 | M5 |
|---|---|---|---|---|---|---|---|---|---|---|
| LIME | 0 | 2 | 0 | 1 | 1 | 0 | 0 | 0 | 0 | 0 |
| RISE | 0 | 0 | 0 | 1 | 1 | 2 | 0 | 0 | 0 | 0 |
| SHAP | 0 | 2 | 0 | 1 | 0 | 1 | 0 | 1 | 1 | 1 |

We note that the opinions of the two radiologists above do not entirely agree with each other - this is due to the fact that identifying all cancerous regions by eye, especially on benign mammograms, is extremely difficult. The scans are also fairly noisy and in parts blurry by nature. The purpose of this form of evaluation was not to have radiologists perfectly highlight all cancerous regions - the goal was to simply analyse their responses to explanations generated by each XAI technique, in order to judge the usefulness of the techniques as diagnostic tools.

## A.8    Explanation Examples

This section contains examples of image explanations as generated by LIME, RISE and SHAP, described in this report. Figure 6 shows LIME explanations, Fig. 7 shows RISE explanations, and Fig. 8 shows SHAP explanations.

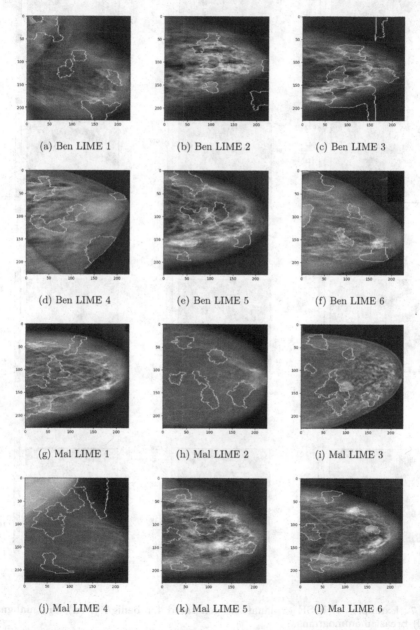

| (a) Ben LIME 1 | (b) Ben LIME 2 | (c) Ben LIME 3 |
| (d) Ben LIME 4 | (e) Ben LIME 5 | (f) Ben LIME 6 |
| (g) Mal LIME 1 | (h) Mal LIME 2 | (i) Mal LIME 3 |
| (j) Mal LIME 4 | (k) Mal LIME 5 | (l) Mal LIME 6 |

**Fig. 6.** Examples of LIME explanations generated for benign (Ben) and malignant (Mal) breast mammograms.

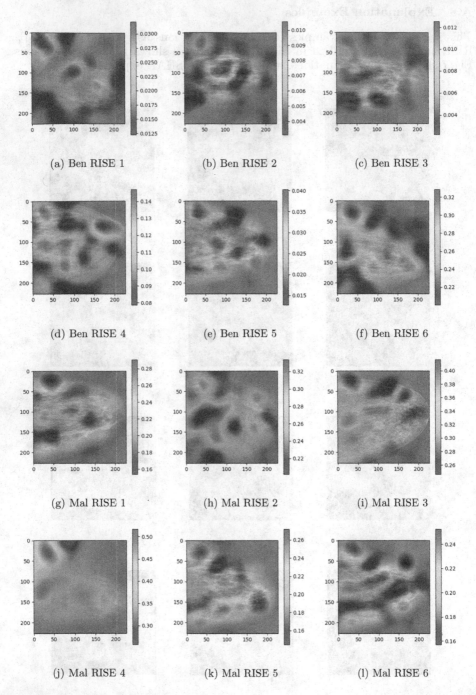

(a) Ben RISE 1     (b) Ben RISE 2     (c) Ben RISE 3

(d) Ben RISE 4     (e) Ben RISE 5     (f) Ben RISE 6

(g) Mal RISE 1     (h) Mal RISE 2     (i) Mal RISE 3

(j) Mal RISE 4     (k) Mal RISE 5     (l) Mal RISE 6

**Fig. 7.** Examples of RISE explanations generated for benign (Ben) and malignant (Mal) breast mammograms.

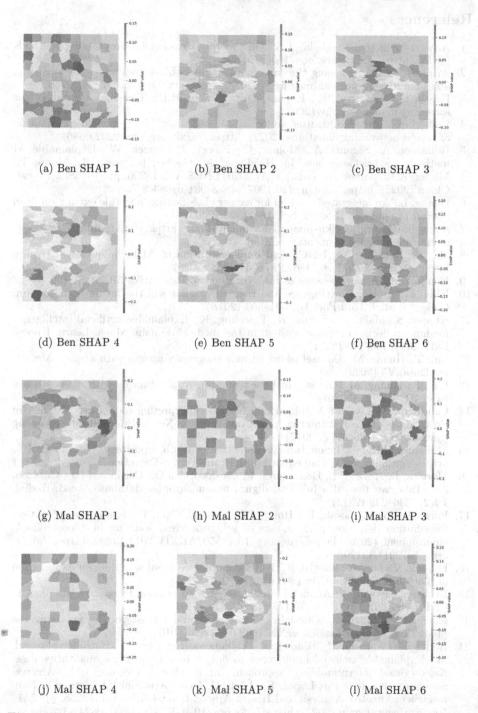

**Fig. 8.** Examples of SHAP explanations generated for benign (Ben) and malignant (Mal) breast mammograms.

# References

1. Alvarez-Melis, D., Jaakkola, T.S.: On the robustness of interpretability methods (2018). https://arxiv.org/abs/1806.08049
2. Arun, N., et al.: Assessing the (un)trustworthiness of saliency maps for localizing abnormalities in medical imaging (2020). https://arxiv.org/abs/2008.02766
3. Heath, M.D., Bowyer, K., Kopans, D.B., Moore, R.H.: The digital database for screening mammography (2007)
4. Hedström, A., et al.: Quantus: an explainable AI toolkit for responsible evaluation of neural network explanations (2022). https://arxiv.org/abs/2202.06861
5. Holzinger, A., Saranti, A., Molnar, C., Biecek, P., Samek, W.: Explainable AI methods - a brief overview. In: Holzinger, A., Goebel, R., Fong, R., Moon, T., Müller, K.R., Samek, W. (eds.) xxAI 2020. LNCS, vol. 13200, pp. 13–38. Springer, Cham (2022). https://doi.org/10.1007/978-3-031-04083-2_2
6. Huang, L.: An integrated method for cancer classification and rule extraction from microarray data. J. Biomed. Sci. **16**(1), 25 (2009)
7. scikit image.org: Scikit-image documentation. https://scikit-image.org/docs/stable/api/skimage.segmentation.html
8. Jia, X., Ren, L., Cai, L.: Clinical implementation of AI techniques will require interpretable AI models. Med. Phys. **47**, 1–4 (2020)
9. Kendall, M.: A new measure of rank correlation. Biometrika **30**, 81–89 (1938)
10. King, B.: Artificial intelligence and radiology: what will the future hold? J. Am. College Radiol. **15**(3 Part B), 501–503 (2018)
11. Knapič, S., Malhi, A., Saluja, R., Främling, K.: Explainable artificial intelligence for human decision support system in the medical domain. Mach. Learn. Knowl. Extr. **3**(3), 740–770 (2021)
12. Lin, T., Huang, M.: Dataset of breast mammography images with masses. Mendeley Data, V5 (2020)
13. Lin, T., Huang, M.: Dataset of breast mammography images with masses. Data Brief **31**, 105928 (2020)
14. Lundberg, S., Lee, S.: A unified approach to interpreting model predictions. In: Proceedings of the 31st International Conference on Neural Information Processing Systems, pp. 4768–4777 (2017)
15. MohamedAliHabib: Brain tumour detection, Github repository. GitHub (2019). https://github.com/MohamedAliHabib/Brain-Tumor-Detection
16. Moreira, I., Amaral, I., Domingues, I., Cardoso, A.J.O., Cardoso, M.J., Cardoso, J.S.: Inbreast: toward a full-field digital mammographic database. Acad. Radiol. **19**(2), 236–248 (2012)
17. Park, J., Jo, K., Gwak, D., Hong, J., Choo, J., Choi, E.: Evaluation of out-of-distribution detection performance of self-supervised learning in a controllable environment (2020). https://doi.org/10.48550/ARXIV.2011.13120. https://arxiv.org/abs/2011.13120
18. Petsiuk, V., Das, A., Saenko, K.: RISE: randomized input sampling for explanation of black-box models. arXiv:1806.07421 (2018)
19. Recht, M., Bryan, R.: Artificial intelligence: threat or boon to radiologists? J. Am. College Radiol. **14**(11), 1476–1480 (2017)
20. Ribeiro, M., Singh, S., Guestrin, C.: "Why should I trust you?": explaining the predictions of any classifier. arXiv:1602.04938v3 (2016)
21. Rodriguez-Sampaio, M., Rincón, M., Valladares-Rodriguez, S., Bachiller-Mayoral, M.: Explainable artificial intelligence to detect breast cancer: a qualitative case-based visual interpretability approach. In: Ferrández Vicente, J.M., Álvarez-Sánchez, J.R., de la Paz López, F., Adeli, H. (eds.) Artificial Intelligence in Neuroscience: Affective Analysis and Health Applications. LNCS, vol. 13258, pp. 557–566. Springer, Cham (2022). https://doi.org/10.1007/978-3-031-06242-1_55
22. Ross, A., Willson, V.L.: One-Way Anova, pp. 21–24. SensePublishers, Rotterdam (2017)

23. Rudin, C.: Stop explaining black box machine learning models for high stakes decisions and use interpretable models instead. Nat. Mach. Intell. **1**, 206–215 (2019)
24. Selvaraju, R., Cogswell, M., Das, A., Vedantam, R., Parikh, D., Batra, D.: Grad-CAM: visual explanations from deep networks via gradient-based localization. arXiv:1610.02391 (2017)
25. Seyedeh, P., Zhaoyi, C., Pablo, R.: Explainable artificial intelligence models using real-world electronic health record data: a systematic scoping review. J. Am. Med. Inform. Assoc. **27**, 1173–1185 (2020)
26. Shrikumar, A., Greenside, P., Kundaje, A.: Learning important features through propagating activation differences (2017). https://arxiv.org/abs/1704.02685
27. Simonyan, K., Vedaldi, A., Zisserman, A.: Deep inside convolutional networks: visualizing image classification models and saliency maps. https://arxiv.org/abs/1312.6034 (2014)
28. Suckling, J., Parker, J., Dance, D.: Mammographic image analysis society (MIAS) database v1.21 (2015). https://www.repository.cam.ac.uk/handle/1810/250394
29. Sun, Y., Chockler, H., Huang, X., Kroening, D.: Explaining image classifiers using statistical fault localization. In: Vedaldi, A., Bischof, H., Brox, T., Frahm, J.-M. (eds.) ECCV 2020. LNCS, vol. 12373, pp. 391–406. Springer, Cham (2020). https://doi.org/10.1007/978-3-030-58604-1_24
30. Sun, Y., Chockler, H., Kroening, D.: Explanations for occluded images. In: International Conference on Computer Vision (ICCV), pp. 1234–1243. IEEE (2021)
31. van der Velden, B.H., Kuijf, H.J., Gilhuijs, K.G., Viergever, M.A.: Explainable artificial intelligence (XAI) in deep learning-based medical image analysis. Med. Image Anal. 102470 (2022)
32. Webber, W., Moffat, A., Zobel, J.: A similarity measure for indefinite rankings. ACM Trans. Inf. Syst. **28**(4), 1–38 (2010)
33. Zeiler, M.D., Fergus, R.: Visualizing and understanding convolutional networks. In: Fleet, D., Pajdla, T., Schiele, B., Tuytelaars, T. (eds.) ECCV 2014. LNCS, vol. 8689, pp. 818–833. Springer, Cham (2014). https://doi.org/10.1007/978-3-319-10590-1_53
34. Zhou, B., Khosla, A., Lapedriza, A., Oliva, A., Torralba, A.: Learning deep features for discriminative localization. arXiv:1512.04150 (2016)

# Author Index

Printed in the United States
by Baker & Taylor Publisher Services